David Carter
435-96-9474
388-3489

Mineral Optics

A Series of Books in Geology

EDITORS: James Gilluly and A. O. Woodford

Mineral Optics

PRINCIPLES AND TECHNIQUES

Wm. Revell Phillips
BRIGHAM YOUNG UNIVERSITY

Illustrations drawn by the author

W. H. FREEMAN AND COMPANY
SAN FRANCISCO

Copyright © 1971 by W. H. Freeman and Company

No part of this book may be reproduced by any mechanical,
photographic, or electronic process, or in the form of
a phonographic recording, nor may it be stored in a
retrieval system, transmitted, or otherwise copied for
public or private use without written permission from the publisher.

Printed in the United States of America

Library of Congress Catalog Card Number: 78-134208
International Standard Book Number: 0-7167-0251-7

2 3 4 5 6 7 8 9 10

Preface

The study of mineral optics has become a standard requirement for students of the earth sciences. It is an essential prerequisite for the study of crystallography, crystal chemistry, and solid-state physics.

This work is designed to prepare students for the study of rocks in thin section (i.e., petrography). The elementary principles of crystal optics are presented to form the basis of a practical approach to the identification of minerals, in fragments and thin sections, by means of the standard petrographic microscope. For a full understanding of the material the student should have a background knowledge of trigonometry, mineralogy (crystallography), and the elementary physics of light. Highly theoretical aspects of optics such as those requiring distinction between wave paths and wave normals, between primary and secondary optic axes, and between ray velocity surfaces and wave velocity surfaces are deliberately de-emphasized or entirely avoided, as they are of little use in the practical problem of mineral identification. Ray velocity surfaces are derived from visual observations and are regarded as real phenomena; from the ray velocity surfaces, the standard indicatrix, which is an imaginary reference surface, is derived for use in representing real phenomena. No other reference surfaces, real or imaginary, are considered.

The last four chapters of this book are concerned with the construction and use of the universal stage—an instrument that is easily understood by students and well suited to their use in the laboratory. Most petrographic laboratories, teaching or research, are now equipped with one or more universal stages. This valuable instrument has proved essential for the accurate measurement of $2V$ and optical orientation, it is indispensable for twinning studies and petrofabrics, and leads to more accurate distinctions of varieties of the feldspars, pyroxenes, amphiboles, olivines, and other complex mineral groups.

The text is well adapted to a fifteen-week (i.e., semester) course of two lectures and two laboratory periods per week or to a ten-week (i.e., quarter) course of three lectures and two laboratory periods per week. For briefer courses one may wish to exclude certain chapters.

For routine mineral identification this volume of theory and techniques should be used in close association with one of several reference works that include tables of optical constants and optical descriptions of common minerals. Such reference works are available on several levels of complexity—from the detailed volumes of Deer, Howie, and Zussman to the "bare necessity" tables of Larsen and Berman. Similar tables of optical constants of inorganic chemical compounds are available in the books listed on page 244.

The writer is indebted to colleagues and friends for suggestions and encouragement.

WM. REVELL PHILLIPS

January 1971

Contents

1. Elementary Concepts of Light   1

   The Nature of Light   1
   Light As a Transverse Wave Phenomenon   3
   Index of Refraction   10
   Polarized Light   13

2. The Petrographic Microscope   17

   Historical Development   17
   Function and Construction   18
   Accessories to the Petrographic Microscope   39
   Proper Use of the Microscope   41
   Adjustment of the Microscope   42

3. Refractometry   47

   General Principles   47
   Immersion Liquids   59

4. Isotropism and Isotropic Media   69

    The Nature of Isotropism   69
    Isotropic Minerals and the Petrographic Microscope   70

5. Optical Crystallography of Uniaxial Crystals   75

    Uniaxial Anisotropism   75
    Uniaxial Ray Velocity Surfaces   79
    The Uniaxial Indicatrix   82
    The Nicol Prism   86

6. Uniaxial Crystals and the Petrographic Microscope   89

    Interference Colors and Birefringence   89
    Uniaxial Interference Figures   103
    Index of Refraction   114
    Color and Pleochroism   115
    Crystal Orientation   116

7. Optical Crystallography of Biaxial Crystals   121

    Biaxial Anisotropism   121
    The Biaxial Indicatrix   121
    Biaxial Ray Velocity Surfaces   129

8. Biaxial Crystals and the Petrographic Microscope   135

    Interference Colors and Birefringence   135
    Biaxial Interference Figures   136
    Index of Refraction   158
    Color and Pleochroism   159
    Crystal Orientation   159

9. The Universal Stage   171

    Function and Construction   171
    Mounting and Adjusting the Universal Stage   177
    Graphical Representation of Measurements   179

10. Application of the Universal Stage to Uniaxial Crystals   191

Distinguishing Uniaxial and Biaxial Minerals   191
Orientation of Uniaxial Crystals   193
Cleavage Studies   196
Twinning Studies   197
Distinguishing Carbonate Minerals in Thin Section   201

11. Application of the Universal Stage to Biaxial Minerals   205

Orientation of Biaxial Minerals   205
Relationships Between Optical and Crystallographic Directions   210
Cleavage Studies   215
Twinning Studies   215
Feldspar Studies   222

12. Preparation of the Sample   231

Mineral Fragments   231
Rock Thin Sections   232

References   237

Index   245

Mineral Optics

CHAPTER 1

Elementary Concepts of Light

The Nature of Light

Light is radiant energy of a wavelength that can stimulate our visual sense. As with electricity and other forms of energy, we know it by its effects, and can predict its behavior and use it without fully understanding its nature.

The early Greeks considered light to be a stream of minute particles, called corpuscles, which either were emitted by the eye or entered the eye from a luminous body. This theory remained virtually unchallenged until the late seventeenth century, when the Dutch scientist Christiaan Huygens (1629–1695), in attempting to explain diffraction, refraction, interference, polarization, and other optical phenomena, proposed that light is propagated as longitudinal waves. But Huygens' contemporary, Sir Isaac Newton (1642–1727), favored the older corpuscular theory, and because of his stature as a scientist the wave concept was largely disregarded until the early nineteenth century. The work of Thomas Young (1773–1829), Augustin J. Fresnel (1788–1827), Jean B. L. Foucault (1819–1868), and others eventually led to the acceptance of the theory that light is a wave phenomenon—but that the waves are transverse rather than longitudinal as Hugyens had suggested. The wave theory of light, either transverse or longitudinal, was still not without major objections, however, for even its foremost defenders were not completely satisfied with its adequacy. Perhaps for them the major problem was that waves require a transmitting medium for their propagation; accordingly, the supporters of the wave theory postulated the existence of an ether—a special medium that fills all space and is endowed with some highly improbable physical properties. The independent discovery by Michael Faraday (1791–1867) and Joseph Henry (1797–1878) of electrical induction and the subsequent discovery by Faraday of the first magneto-optic phenomenon were instrumental in drawing the attention of James Clerk Maxwell (1831–1879) to the possible connection between light, electricity, and magnetism. Through his electromagnetic theory Maxwell showed that electricity and magnetism were inseparable phenomena. He concluded that if his theory were correct, electromagnetic waves must travel at the velocity of light, and, furthermore, that light itself must be an electromagnetic phenomenon. It was Heinrich Hertz (1857–1894) who provided experimental confirmation of Maxwell's theory by demonstrating that electromagnetic disturbances have measurable wavelengths and generally possess the properties of light waves.

For a while it seemed that Hertz's work represented the last chapter to be written on the nature of light, but with the beginning of the present century came the dawn of a new era in the physical sciences—an era in which the work of such intellectual giants as J. J. Thomson (1856–1940), Ernest Rutherford (1871–1937), Niels Bohr (1885–1962), Henry Moseley (1887–1915), Max Planck (1858–1947), and Albert Einstein (1879–1955) gradually revealed the inner world of the atom. Besides being the first to propose a theory of atomic structure, Thomson also introduced the notion that electricity is of a corpuscular nature. For several decades the world of physics was divided over whether various kinds of radiation were wave phenomena or corpuscular phenomena. Planck's work on black-body radiation led him to postulate, in 1900, that a light source emits its radiation in discrete quanta, or particles, rather than continuously as the electromagnetic theory held. The quantum theory was so revolutionary, however, that it was generally disregarded until Albert Einstein made use of it in 1905 to explain the photoelectric effect— a phenomenon that Hertz had discovered before the turn of the century but which had never been explained in terms of the electromagnetic theory.

Physicists must still resort to using two seemingly contradictory theories to explain various luminous phenomena. Although the quantum theory and wave mechanics have supplied explanations for certain phenomena that could not be explained in terms of the electromagnetic theory, they did not lead to the development of a unified theory of light. Some phenomena can be explained only in terms of the wave theory, and others only in terms of the quantum concept (see Table 1-1). The two theories may soon be reconciled, however,

TABLE 1-1. The Different Theories of Light and the Groups of Phenomena to Which They Pertain

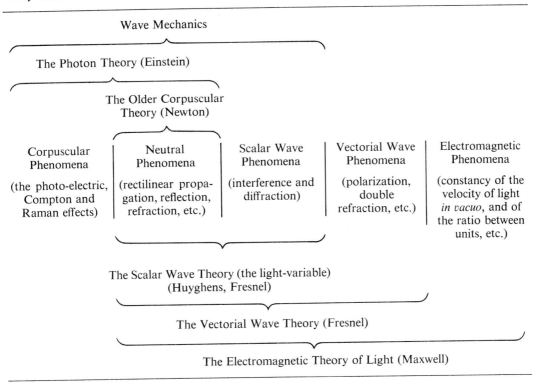

Source: Louis De Broglie, Matter and Light: The New Physics (translated by W. H. Johnston), W. W. Norton & Company, Inc., 1939.

for the equivalence of matter and energy has been demonstrated repeatedly. Moreover, not only do light quanta (photons) behave like particles of matter, but electrons behave like groups of waves.

Light As a Transverse Wave Phenomenon

Wave Motion

Since in the study of light behavior within crystalline media we are concerned chiefly with refraction, polarization, and interference, we shall consider light to consist of transverse waves transmitted by atomic vibration.

The simplest form of wave is that produced by simple harmonic motion. Such a wave is shown in Figure 1-1; the direction of propagation is from left to right along the x-axis, and the direction of vibration is perpendicular

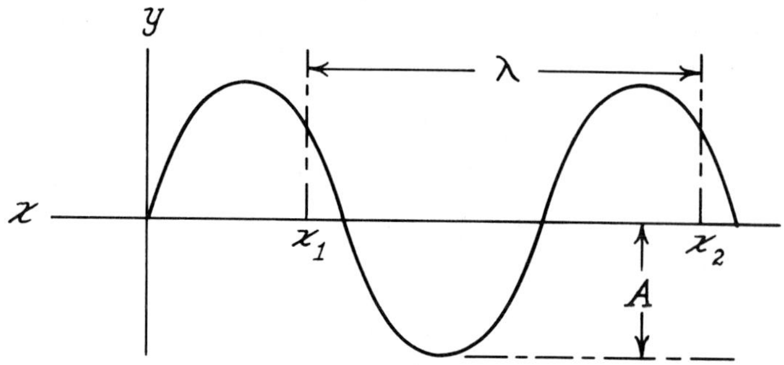

FIGURE 1-1. A wave produced by simple harmonic motion.

to the line of propagation (i.e., in the plane of the page). A light wave causes particles to vibrate in planes perpendicular to the direction of propagation *without forward motion*, just as water molecules are caused to vibrate up and down when a wave crosses a pond. Simple harmonic motion can be expressed mathematically as

$$y = A \sin x,$$

where A is the amplitude of the wave, or the maximum displacement of any particle from the line of propagation. The amplitude of the wave is a measure of light intensity, and the rate of energy propagation is proportional to the square of the amplitude.

The distance between two successive wave crests or troughs, or between any two corresponding points on successive waves ($x_2 - x_1$), is the wavelength (λ), which is intimately related to color. Color is actually a function of the number of vibrations per second that enter the eye, or in other words, the frequency (f) of the light, which is related to the wavelength by the formulas

$$v = \lambda f \quad \text{and} \quad v = \lambda/T,$$

where T is the period of vibration (i.e., the time required for one complete vibration), or the time represented by ($x_2 - x_1$), and v is the velocity of the light. When light passes from one transparent medium to another of different optical density, both velocity and wavelength change but frequency remains unchanged.

Velocity of Light

The velocity of light defied measurement until 1675, when Olaus Roemer (1644–1710) noticed that eclipses of the moons of Jupiter came progressively earlier than calculated as the earth approached Jupiter in its orbital path, and progressively later as the earth receded. The eclipses were more than twenty seconds later than predicted when the light from Jupiter travelled the additional 186,000,000 miles across the earth's orbit. Assuming that light must have a finite velocity, Roemer proposed a value of about 141,000 miles/sec for the velocity of light. The first measurements not based on astronomical observations were made in 1849 by Armand H. L. Fizeau, who used a rapidly rotating wheel of mirrors reflecting light from a distant mirror. His value was about five percent too large. Fizeau's procedure and results were successively refined by Jean B. L. Foucault in 1850 and by Albert A. Michelson in 1882, 1927, and 1933. The presently accepted value of the velocity of light is 2.997925×10^{10} cm/sec, or 186,282 miles/sec, in free space.

Although the velocity of light in free space is one of the most accurately known of the physical constants, the velocity of light in general is certainly not constant. The velocity of light in all media other than free space (i.e., a vacuum) depends both on the nature of the medium and the frequency of the light.

Phase

Two or more waves moving along the same line of propagation and polarized to vibrate in the same plane will interfere, since the displacement of a given particle of the transmitting medium may be either aided (constructive interference) or opposed (destructive interference) by the second wave. The resultant of two waves is obtained by addition (Fig. 1-2,A). If the distance between corresponding points on two waves of equal amplitude is some whole multiple n of the wavelength ($n\lambda$), the two waves are said to be "in phase," and reinforcement is maximal (i.e., the amplitude of the resultant is twice that of the component waves, as in Fig. 1-2,B). Otherwise, the waves are "out of phase." The most that two waves can be out of phase is when the distance between corresponding points is $n + \frac{1}{2}\lambda$, which results in complete annulment (i.e., the resultant is a straight line, as is Fig. 1-2C).

The Ray-Velocity Surface

A light ray may be considered as a line segment beginning at the point of origin of a wave and increasing in length as the wave advances along the direction of propagation. A pencil of light is a small bundle of slightly diverging rays originating from a point source, and a beam of light is a small bundle of strictly parallel light rays.

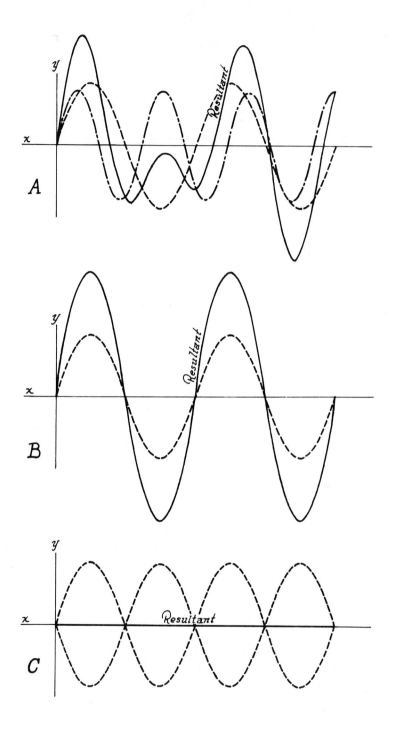

FIGURE 1-2. Wave addition. (A) The addition of two waves moving along the same line of propagation and lying in the same plane is determined by the arithmetic sum of the two forces acting on each transmitting particle. (B) If the two waves are of equal amplitude and differ in phase by $n\lambda$, the force acting on each transmitting particle is twice that of a single wave, and the resultant has twice the amplitude of its component waves. (C) If the two waves are of equal amplitude and differ in phase by $(n + \frac{1}{2}\lambda)$, forces acting on every transmitting particle are equal and opposite and the resultant wave has zero amplitude.

Let us now consider the light emitted from a point source. Rays move outward in all directions from the source and, since light travels with finite velocity all rays will have travelled a certain distance after any given time. The ray velocity surface is the surface formed by the ends of all advancing rays at any instant of time. Since an *isotropic medium* is defined as one in which light travels with equal velocity in all directions, the ray velocity surface within any isotropic medium is a sphere that grows larger as the rays advance. In this book the emphasis is mainly on *anisotropic media*, in which light velocity is a function of the direction of propagation and in which the ray velocity surface is not a simple sphere.

The Electromagnetic Spectrum

Visible light is but a small part of a continuous series of electromagnetic radiation ranging from more than 1,000 km in wavelength to less than 10^{-10} cm. The electromagnetic spectrum is divided into regions on the basis of both application and the mechanism used to detect the radiation (Fig. 1-3). The regions are unequal in range, and their limits are arbitrary and frequently overlapping. Visible light is only that part of the series which stimulates our eyes. It ranges in wavelength from about 7.6×10^{-5} cm to about 4×10^{-5} cm. Several dimensional units are commonly used for these shorter wavelengths. A micron (μ) is 0.001 mm, or 10^{-4} cm; a millimicron (mμ) is consequently 10^{-7} cm, and an Angstrom unit (Å) is 10^{-8} cm.

Dispersion

The sun's light, called white light, contains essentially all wavelengths of the visible range. This is familiar to anyone who has ever observed a rainbow or used a glass prism to spread the sun's light into its component colors (Fig. 1-4). The separation of *polychromatic light* (i.e., light that consists of more than one wavelength) into its component wavelengths is called dispersion.

FIGURE 1-3. The electromagnetic spectrum. The spectrum is divided into designated regions on the basis of use and mode of detection. The limits of the regions are often indefinite, and ranges often overlap.

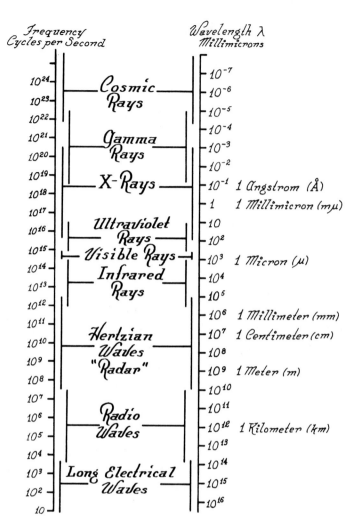

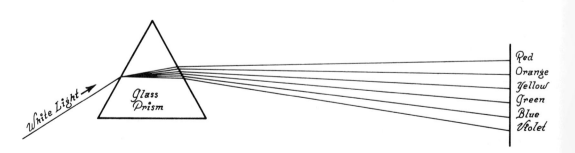

FIGURE 1-4. Dispersion by a simple glass prism.

Color

Color, like light itself, is simply our perception of visual stimuli. Each wavelength of light produces a different color sensation, and although *monochromatic light* (i.e., light consisting of only one wavelength) is never truly attained in practice, it is virtually produced by several light sources. Our eyes, however, cannot distinguish whether the light they receive is monochromatic or consists of certain wavelength combinations that produce a color sensation indistinguishable from monochromatic radiation.

LIGHT ADDITION. White light is produced by all colors acting simultaneously on our eyes, but may also be produced by specific combinations of only two or three colors. Any two colors of light that combine to yield white light are *complementary light colors*. Red and blue-green, yellow and blue-violet, and green and violet are complementary light colors. Red, green, and blue-violet are the only three colors that combine to form white light, and when combined in proper proportions are capable of producing light of any color. These are called the *primary light colors*. The process of the addition of light colors was well demonstrated by Von Nardoff, who projected on a screen partially overlapping circles of red, green, and blue-violet light (Plate I; the plates follow page 148). All three colors overlap to yield white light, and any two primary colors overlap to form the complementary color of the third primary.

LIGHT SUBTRACTION. Light incident on an object is partially reflected, partially absorbed, and partially transmitted. The color of a transparent object is the color of the light transmitted though it, and that of an opaque object is the color of the reflected light, the remainder of the light being absorbed. Pigments are substances added to paints, inks, etc., to produce colors. The pigments cause certain incident wavelengths to be subtracted, or absorbed. A black object absorbs all incident light. A white object reflects all colors, and an object that partially absorbs and reflects all colors appears gray. A "white" object in red light is red, since only red light can be reflected. The color of an object is thus not an inherent property of the object, but a function of the light received by our eyes from the object. The *primary pigment colors*, red, blue, and yellow, may be mixed to yield black and, in proper proportions, any other color. The *complementary pigment colors* are any two that may be mixed to yield black; they are red and green, yellow and violet, and blue and orange. Thus each set of complementary colors consists of one primary and one secondary color resulting from the combination of the other two primary colors.

Index of Refraction

Definition

A familiar law of physics states that light travels in a straight line. This is of course true, but only if the light rays do not pass from one medium into another of different optical density. Should this happen, the light rays are bent, or refracted, and no longer follow their original path. Refraction is the result of a difference in the velocity of light in the two media. The quantitative measure of refraction is a pure number called the index of refraction (n), which for any medium is defined as the ratio of the velocity of light in a vacuum (v) to its velocity in that medium (V):

$$n = v/V.$$

The index of refraction of a vacuum is therefore unity by definition. The velocity of light in air is slightly less than in free space because of the density of air; the refractive index of air at atmospheric pressure is 1.000274 at 15°C. This value is so close to unity, however, that it is here considered such.

Consider two parallel light rays passing from one medium (index of refraction n_1) into a second medium (index n_2) (Fig. 1-5). On entering the second medium, the rays are refracted toward the normal to the interface. At a given instant, ray R_1 reaches the interface at o', and R_2 is at point a. In a time interval Δt, ray R_1 moves from o' to b, and R_2 advances from a to o. Since velocity is distance divided by time, the velocity in the first medium (V_1) and the velocity in the second (V_2) are given by

$$V_1 = \frac{ao}{\Delta t} \quad \text{and} \quad V_2 = \frac{bo'}{\Delta t}.$$

Thus

$$\frac{V_1}{V_2} = \frac{ao}{bo'}.$$

From a consideration of perpendicular sides we see that $\angle oo'a = i$ and $\angle o'ob = r$. Therefore,

$$\sin i = \frac{ao}{oo'} \quad \text{and} \quad \sin r = \frac{bo'}{oo'}.$$

Moreover,

$$\frac{\sin i}{\sin r} = \frac{ao}{bo'} \quad \text{or} \quad \frac{V_1}{V_2} = \frac{\sin i}{\sin r},$$

which is Snell's Law.

If

$$n_1 = \frac{v}{V_1} \quad \text{and} \quad n_2 = \frac{v}{V_2},$$

then

$$\frac{n_2}{n_1} = \frac{\sin i}{\sin r}.$$

If the first medium is a vacuum (or air), n_1 is unity and

$$n_2 = \frac{\sin i}{\sin r}.$$

If n_2 is greater than n_1 the angle of incidence (i) is greater than the angle of refraction (r), and if n_1 is greater than n_2 it is less than r. We may conclude, therefore, that *a light ray passing from one medium into another of higher index is refracted toward the normal to the interface, and conversely, that a light ray passing from one medium into another of lower index is refracted away from the normal.*

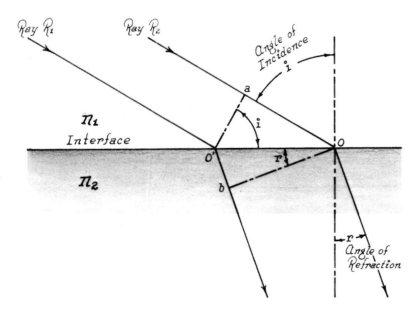

FIGURE 1-5. Refraction. Light rays passing from a medium of low refractive index (n_1) into one of higher refractive index (n_2) are refracted toward the normal to the interface. The angle of incidence (i) and angle of refraction (r) are related by the equation $\sin i/\sin r = n_2/n_1$.

Total Reflection and Critical Angle

If the light rays shown in Figure 1-5 were reversed, the angle of incidence would become the angle of refraction, and vice versa (Fig. 1-6). Since the indices of the media do not change,

$$\frac{n_2}{n_1} = \frac{\sin r}{\sin i}.$$

The light ray R_1 is refracted away from the normal as it enters the medium of lower index; if its angle of incidence is increased, the angle of refraction also increases until a path R_2 is reached, where the ray is refracted parallel to the interface. If the angle of incidence is further increased, as in path R_3, the light ray undergoes *total reflection;* that is, the ray is completely reflected back into the medium. The angle of reflection (r_3') is of course equal to the angle of incidence (i_3). For the ray path R_2, whose angle of refraction is 90°, the angle of incidence is the *critical angle* (CA).

$$\frac{n_2}{n_1} = \frac{\sin (90°)}{\sin (CA)} = \frac{1}{\sin (CA)}.$$

If the low-index medium is air, $n_1 = 1$, and

$$\sin (CA) = \frac{1}{n_2} \quad \text{or} \quad n_2 = \frac{1}{\sin (CA)}.$$

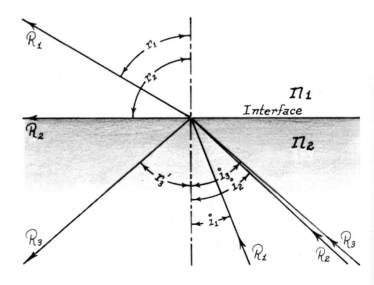

FIGURE 1-6. Total reflection and critical angle. Light rays passing from a medium of high refractive index (n_2) into a medium of lower refractive index (n_1) are totally reflected at the interface when the angle of incidence exceeds the critical angle (i_2).

Polarized Light

According to the transverse wave theory, ordinary light is composed of waves vibrating in all possible planes through the line of propagation (Fig. 1-7,A). Polarized light waves vibrate in only one, or two, planes and thus do not vibrate symmetrically about the direction of propagation (Fig. 1-7,B). This is our best reason for accepting the transverse wave theory, since it is impossible to imagine a longitudinal wave or a stream of corpuscles that would not be symmetrical about the line of propagation. If we assume transverse waves, we must assume a transmitting ether and, since an ether particle cannot vibrate in all directions simultaneously, ordinary light must consist of an infinite number of polarized waves, each vibrating in different planes in such rapid succession that all directions of vibration about the line of transmission are equally represented.

Plane polarized light vibrates in a single plane called the plane of polarization. According to Maxwell's electromagnetic theory, a light wave consists of an electric wave and a magnetic wave vibrating in mutually perpendicular planes that intersect and form an axis along the line of transmission. The plane of vibration of the light wave is that of the electric wave. The term "plane of polarization" is often used to denote the plane of a magnetic wave,

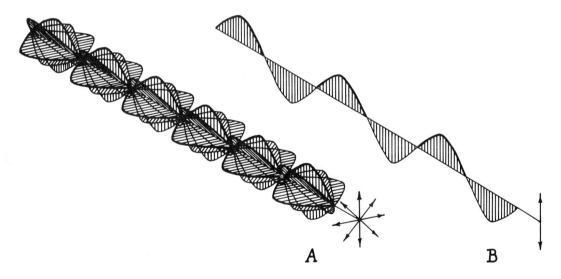

FIGURE 1-7. Plane polarized light. (A) Ordinary light is a combination of waves vibrating in all possible planes through the line of propagation. (B) Plane polarized light is represented by waves confined to vibration in only one plane through the line of propagation.

and hence the plane of polarization is perpendicular to that of vibration. Commonly, however, the term "polarization plane" and "vibration plane" are used interchangeably. This is done here in the hope of lessening confusion.

Polarization by Reflection

In 1882 Malus discovered that light reflected at certain angles from a smooth surface of a transparent, nonconducting substance is plane polarized, with its plane of maximum polarization parallel to the reflecting surface (Fig. 1-8). Sir David Brewster demonstrated that complete polarization of the reflected beam occurs when the angle of incidence and the angle of refraction total 90° (Fig. 1-9). The angle of incidence at which this condition obtains is called the polarizing angle (p). If n_1 is the index of refraction of air (or a vacuum) and n_2 the index of the refracting medium, then

$$\frac{\sin i}{\sin r} = \frac{n_2}{n_1} = n_2 = \frac{\sin p}{\sin (90° - p)} = \frac{\sin p}{\cos p} = \tan p.$$

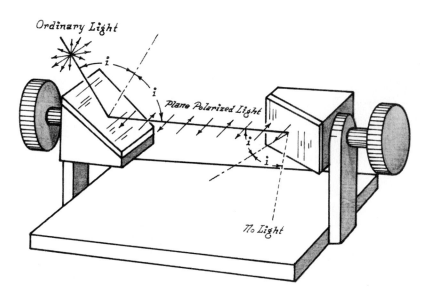

FIGURE 1-8. Polarization by reflection. Ordinary light is partially polarized by reflection from the smooth surface of a transparent, nonconducting medium. The plane of polarization of the reflected light is parallel to the reflecting surface. For ordinary glass, maximum polarization occurs when the angle of incidence is about 57°.

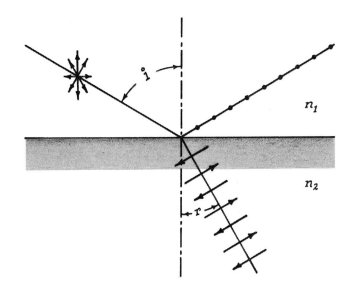

FIGURE 1-9. Brewster's Law. Complete polarization of a reflected beam occurs when $i + r = 90°$. The angle of incidence (i) is then called the polarizing angle (p) and, by Brewster's Law, $\tan p = n_2/n_1$. Note that reflected waves vibrate parallel to the reflecting surface and that refracted waves are polarized (in part) normal to the reflected waves.

Brewster's Law states accordingly that the tangent of the polarizing angle is equal to the index of refraction of the refracting medium.

Not only is the reflected beam completely polarized parallel to the reflecting surface, but the refracted beam is partially polarized parallel to the plane of the rays. Since only a little incident light is reflected, a pile of six or eight reflecting plates can be used to increase the intensity of the polarized beam by adding the rays reflected from all surfaces, a method that was used extensively in some early polarizing instruments.

Polarization by Absorption

Some transparent crystalline compounds transmit light waves vibrating in one direction while strongly absorbing those vibrating in all others—a property of an *anisotropic* crystal structure. A plate of tourmaline cut parallel to the c-crystallographic axis transmits light vibrating parallel to c and strongly absorbs components perpendicular to c. When two such crystal plates are *crossed*—that is, when the axis of one is oriented perpendicular to that of the other—the light transmitted by one crystal is strongly absorbed by the other, and virtually no light is transmitted where the two crystals overlap (Fig. 1-10). If one crystal is slowly rotated, however, light in the overlapping zone increases from a minimum when the c-axes are perpendicular, to a maximum when parallel.

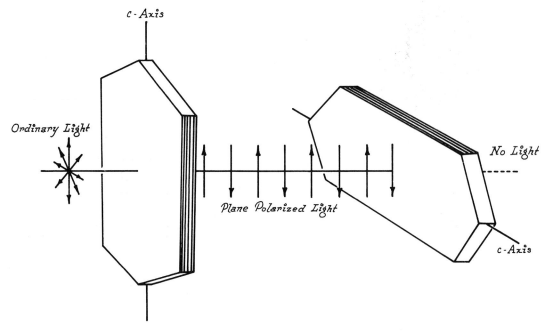

FIGURE 1-10. Polarization by absorption. Tourmaline crystals strongly absorb light waves vibrating perpendicular to the *c*-crystallographic axis and transmit waves vibrating parallel to *c*.

In 1852, William Herapath described crystals of an organic compound that he called iodocinchondine-sulfate, which, like tourmaline, polarizes light by absorption. The compound was subsequently named herapathite in his honor. Myriads of these tiny slender crystals oriented in a uniform direction in a plastic binder are the basis of the original commercial "Polaroid," which was developed in the 1930's by Edwin H. Land. The uniform orientation of microscopic herapathite crystals was achieved by mixing them in a viscous plastic and extruding the mixture through a thin slit. Today the name "Polaroid" refers to several synthetic filters produced by the Polaroid Corporation of America. These polarizing filters consist of films of long-chain polymers oriented in a uniform direction by stretching and chemical treatment to form highly absorbing crystalline compounds, such as polyvinylene. Polaroid filters in the form of plastic sheets or plates are now available and have largely replaced other materials used for polarizing light. The better polaroids are as effective as any other polarizers, and are generally cheaper and much more convenient.

CHAPTER 2

The Petrographic Microscope

Historical Development

Although the microscope was a well-known scientific instrument before the nineteenth century, and had been widely applied in many branches of science, geologists scorned the instrument on the grounds that one doesn't look at mountains with a microscope. In 1829, the *Edinburgh New Philosophical Journal* published a two-page article entitled "The Nicol Prism" by William Nicol, (1768–1851), a lecturer in natural philosophy at Edinburgh. The prism was made of two pieces of calcite cemented with Canada balsam in such a way as to produce plane polarized light. Two years later Nicol published a second article in which he outlined the steps in the preparation of thin sections of minerals and fossil wood for microscopic examination. With these two articles, William Nicol presented to geology the basic tools of modern petrography.

Not until after Nicol's death, however, did his highly valued but essentially useless collection of optical instruments and thin sections inspire Henry Clifton Sorby (1826–1908), a young student at Edinburgh, to champion the use of the microscope for the study of rocks. Sorby's voluminous writings, published in the 1850's and 1860's, were poorly received in his own country, but they interested several investigators on the continent, notably, Zirkel, Vogelsang, and

Rosenbusch in Germany and Fouqué and Michel-Lévy in France. Within a few decades, work done with the microscope had elevated petrography to the status of a well-accepted and flourishing branch of the natural sciences.

Function and Construction

The petrographic microscope is a modification of the compound microscope commonly used in laboratories. Unlike the simple, single-lens microscopes made by Janssens, Galileo, and others, a compound microscope has two lenses: an objective and an ocular, or eyepiece (Fig. 2-1). The object to be viewed is placed just outside the principal focus (f_1) of the objective lens, which forms a real image, called the principal image, between the ocular and its principal focus (f_2). The ocular, in turn, forms an imaginary image of the real image; it is this imaginary image that we see. The total magnification of the microscope is the product of the magnification of objective and ocular:

$$M_{\text{total}} = M_{\text{objective}} \times M_{\text{ocular}}.$$

The magnification of most petrographic microscopes ranges from about $30\times$ to $500\times$.

The modifications that render the petrographic microscope suitable for detailed study of the optical behavior of transparent crystalline substances are a rotating stage, an upper polarizer (called the analyzer), a lower polarizer, and a Bertrand lens. Other, independent optical accessories are also used in conjunction with the petrographic microscope.

The Microscope Tube

In its simplest form, the microscope tube is a straight metal tube that separates objective and ocular (Fig. 2-2,A). The simple tube of most student microscopes

FIGURE 2-1. Image formation by the compound microscope. The location and size of the image formed by a lens can be obtained graphically by constructing pairs of rays from the outer edges of the object. One of each pair is drawn from the object to the lens and parallel with the lens axis. These rays will be refracted by the lens and pass through the focal point. The rays drawn through the center point of the lens are unrefracted and continue in straight lines. Where the two pairs of rays intersect, the object points (A_1, B_1) are focused as image points (A_2, B_2). The objective lens forms an enlarged, real image (A_2B_2) inside the focal length of the ocular and beyond the focal length of the objective. The ocular forms an enlarged, virtual image (A_3B_3) of the real image (A_2B_2). The lens of the eye focuses the enlarged, virtual image on the retina at A_4B_4.

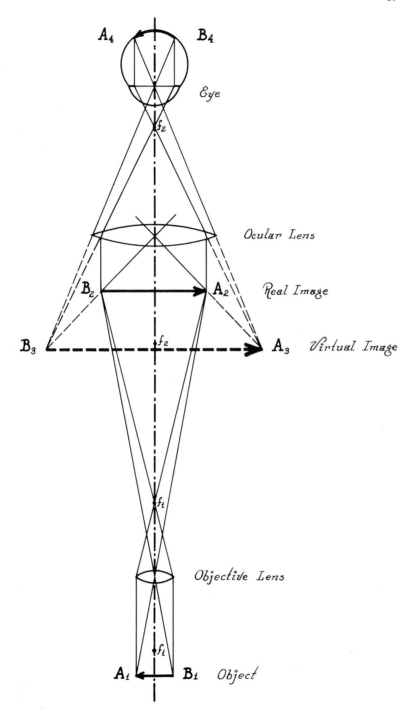

can be detached from the microscope body by raising the tube to its upper limit with the coarse focusing adjustment and lifting it upward. In microscopes of more complex design, the convenience of a horizontal stage is combined with the comfort of an inclined ocular by means of an inclined microscope tube (Fig. 2-3,*A*). A deflecting prism built in at the bend produces slight polarization, which must be compensated for by placing a birefringent plate below the prism, and may also reverse the visual image. Microscopes of high quality contain additional lens elements (Telan lenses) in the microscope tube to make light rays parallel in the vicinity of accessory plates to eliminate image distortion and displacement by accessory plates.

The Ocular, or Eyepiece

The removable lens assembly that fits into the top of the microscope tube is the ocular (Figs. 2-2,*B* and 2-3,*B*). The assembly consists of two plano-convex lenses; the small upper lens is called the eye lens and the large lower lens the field lens. The lenses of the common Huygenian ocular (Fig. 2-6,A) are both made of the same kind of glass, are separated by about half the sum of their focal lengths, and are mounted with both planar surfaces facing the eye of the observer; this combination satisfies the requirements for minimum spherical abberation. The lenses of the Ramsden ocular (Fig. 2-6,B) are of equal focal length, are separated by about one-third the sum of their focal lengths, and are oriented so that their convex surfaces face one another. Although Ramsden oculars are not designed for minimum spherical aberration (i.e., they are not achromatic), they are preferred for quantitative measurements and photomicroscopy, because both image and cross hairs are seen through both lenses and are equally distorted. The Huygenian ocular is said to be a negative ocular, because its focal plane lies between eye lens and field lens; the Ramsden ocular is positive, because its focal plane lies below the field lens. At the focal plane of either type of ocular is a fixed diaphragm that marks the limits of the visible field and a set of cross hairs composed of either delicate wires or lines engraved on a glass disk. Principal image, diaphragm, cross hairs, and dust on the glass disk should all be simultaneously in focus, since they all lie in the focal plane of the ocular. Rotation of the upper collar of the ocular raises or lowers the eye lens to bring cross hairs and diaphragm into sharp focus. Two small slots on the upper edge of the microscope tube are shaped to receive a small projection of the ocular collar to orient the cross hairs either north-south and east-west* or 45° to these conventional directions.

---

*By convention we describe the field of view in terms of map directions: the top of the field is north, the bottom south, and left and right are west and east, respectively.

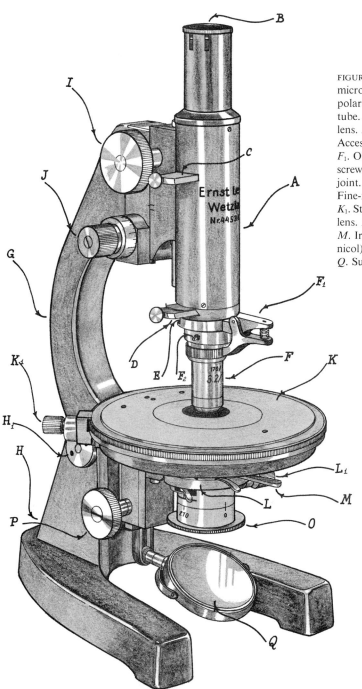

FIGURE 2-2. Vertical-tube petrographic microscope (E. Leitz, Inc.; IIIM polarizing microscope). A. Microscope tube. B. Ocular (eyepiece). C. Bertrand lens. D. Analyzer (upper nicol). E. Accessory opening. F. Objective lens. F_1. Objective clutch arm. F_2. Centering screws. G. Arm. H. Base. H_1. Inclination joint. I. Coarse-focus adjustment. J. Fine-focus adjustment. K. Rotating stage. K_1. Stage clamp. L. Permanent condensing lens. L_1. Auxiliary condensing lens. M. Iris diaphragm. O. Polarizer (lower nicol). P. Substage height adjustment. Q. Substage mirror.

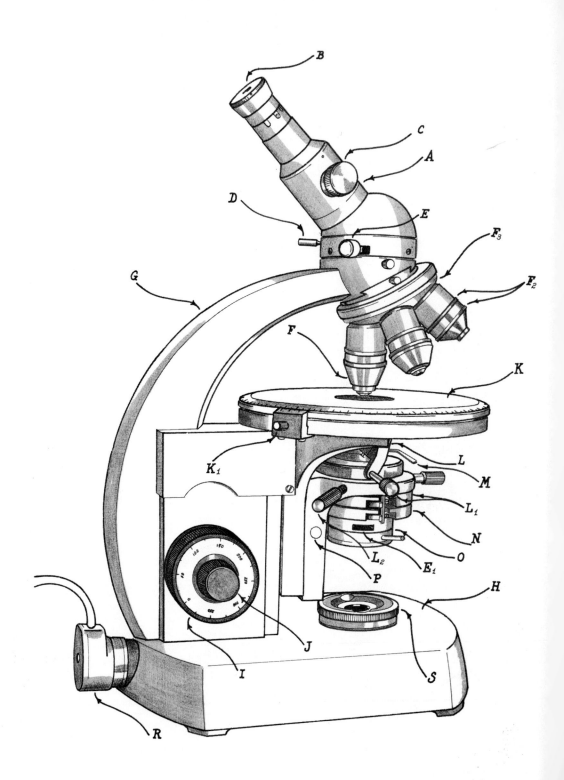

Magnification of the ocular is usually marked on the upper metal surface and ranges from about 2× to 16×. Some oculars are also marked with a field-of-view number that gives the visible field diameter in millimeters when divided by the magnification of the objective with which it is used.

A glass disk engraved with equally spaced parallel lines or with a grid pattern may be placed on the field diaphragm for use in grain-size determination and other quantitative work. A stage micrometer (Fig. 2-7) consisting of a glass plate, or slide, on which are engraved parallel lines of known spacing, usually 0.01 mm, is also used for making linear measurements. The student is encouraged to determine the diameter of his field of view for each ocular-objective combination and to make a notebook drawing of the divisions of a stage micrometer as they appear for each magnification so that he may easily estimate the size of mineral grains (Fig. 2-8). The field diameter of most student-model microscopes is about 1.5 to 2.0 mm at 100× magnification.

The Bertrand Lens

The Bertrand lens—more accurately, the Amici-Bertrand lens—is a small achromatic doublet that can be inserted into the optical path and removed from it by moving the lens horizontally in and out of the tube on a dove-tail slider or, in some microscopes, by rotating the lens mount within the microscope tube (Figs. 2-2,C and 2-3,C). The lens mount can usually be removed for inspection and cleaning by unscrewing a pull knob or a few small screws on the rotating mount. In combination with the ocular the Bertrand lens forms a low-power microscope focused on the upper surface of the objective lens and is used to observe light patterns (i.e., interference figures) formed on this surface. Interference figures can be viewed without the aid of a Bertrand lens by removing the ocular and looking directly down the microscope tube onto the upper surface of the objective. This procedure produces a smaller, but often clearer, pattern.

Some research microscopes have adjustments for centering and focusing the Bertrand lens plus an iris diaphram for isolating a figure formed by a single small crystal grain.

FIGURE 2-3. Inclined-tube petrographic microscope (Carl Zeiss Inc.; Standard RP-48 Pol). *A*. Microscope tube. *B*. Ocular (eyepiece). *C*. Bertrand lens. *D*. Analyzer (upper nicol). *E*. Accessory opening (upper). E_1. Accessory opening (lower). *F*. Objective lens. F_2. Centering collars. F_3. Nose tureet. *G*. Arm. *H*. Base. *I*. Coarse-focus adjustment. *J*. Fine-focus adjustment. *K*. Rotating stage. K_1. Stage clamp. *L*. Permanent condensing lens. L_1. Auxilary condensing lenses. L_2. Centering screws for condensing system. *M*. Iris diaphragm. *N*. Filter holder. *O*. Polarizer (lower nicol). *P*. Substage height adjustment (opposite side). *R*. Illuminator. *S*. Iris diaphragm (illuminator).

A

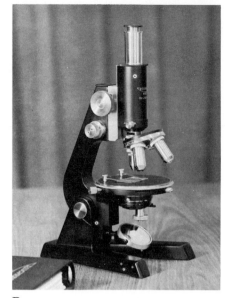
B

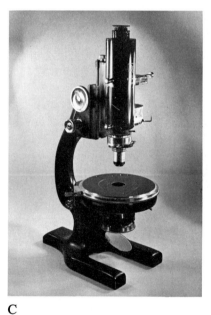
C

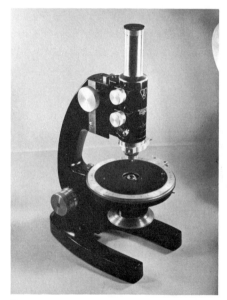
D

FIGURE 2-4. Vertical-tube, student-model petrographic microscopes. Although many models of this type are no longer manufactured, these microscopes are still very common in university classrooms. (A) Unitron; polarizing microscope MPS. (B) C. Reichert; polarizing microscope RCP. (C) American Optical Co.; AO Spencer polarizing microscope. (D) Bausch and Lomb Optical Co.; polarizing microscope L1-2.

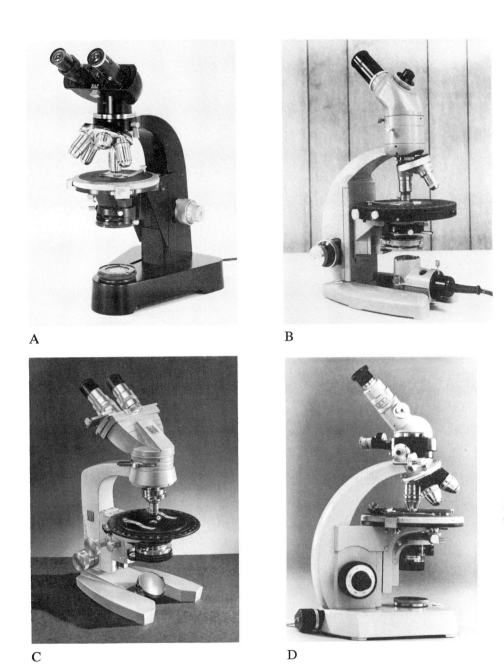

FIGURE 2-5. Research petrographic microscopes. Most petrographic microscopes manufactured today have inclined tubes and are focused by raising or lowering the stage. (A) E. Leitz Inc.; Lobolux pol. (B) Vickers Instruments Inc.; Standard M70A. (C) American Optical Co.; Polarstar 2300 BC-QR. (D) Carl Zeiss Inc.; RA Pol.

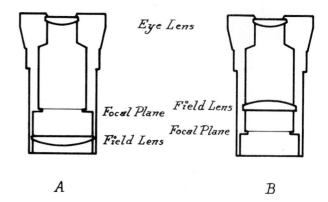

FIGURE 2-6. Positive and negative oculars. (A) Negative (Huygens) ocular. Field diaphram and cross hairs lie in the focal plane of the ocular—between eye lens and field lens. (B) Positive (Ramsden) ocular. Field diaphram and cross hairs lie in the focal plane of the ocular—below the field lens.

The Analyzer, or Upper Nicol

The polarizing device that can be inserted between the Bertrand and objective lenses is called the analyzer, or upper nicol (Figs. 2-2,*D* and 2-3,*D*). The term "nicol," of course, refers to the original calcite polarizer of William Nicol but is here used with reference to any polarizing device made for use with the petrographic microscope. In most student microscopes, the analyzer is fixed to pass only the light wave components that vibrate in an east-west plane and is used in combination with the polarizer, or lower nicol. Since the polarizer passes only north-south waves and the analyzer only east-west waves, the presence of both nicols in the light path (i.e., "crossed nicols") should produce a totally black field of view. In some microscopes the convention is reversed, so that the polarization plane for the analyzer is north-south and the polarization plane for the polarizer is east-west. In many high-quality microscopes the polarizers can be adjusted to suit the preference of the user.

In most late-model polarizing microscopes polarizing filters are used in place of the bulky calcite prisms because they take up less space in the microscope tube and are much cheaper. The filters are made of polyvinyl films that show no selective color absorption and require no compensating lenses. Some research microscopes—those used for work where uniform absorption is required beyond the visible range—are still equipped with calcite prisms, of which there are three types: Nicol, Glan-Thompson, and Ahrens (Fig. 2-9).

The design of the original Nicol prism (described on page 86) takes advantage of natural rhombohedral forms to avoid waste of expensive optical calcite, but the inclined upper and lower surfaces of the prism cause image displacement and loss of illumination by reflection. Although the Glan-Thompson and Ahrens prisms are both of wasteful design, they do reduce these defects to a minimum. The introduction of a calcite analyzer into the light path causes slight shifts in the position of the primary image, making refocusing necessary unless the microscope is equipped with correcting lenses above and below the analyzer. In all research microscopes the analyzer can be rotated to allow observation with "parallel nicols."

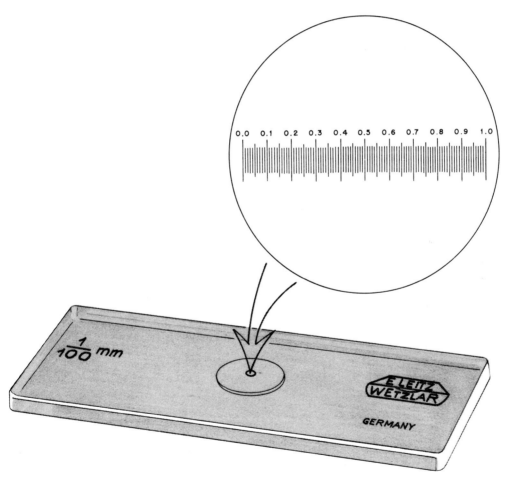

FIGURE 2-7. Stage micrometer. The scale of the micrometer is usually marked in 0.01-mm subdivisions for measuring microscopic dimensions.

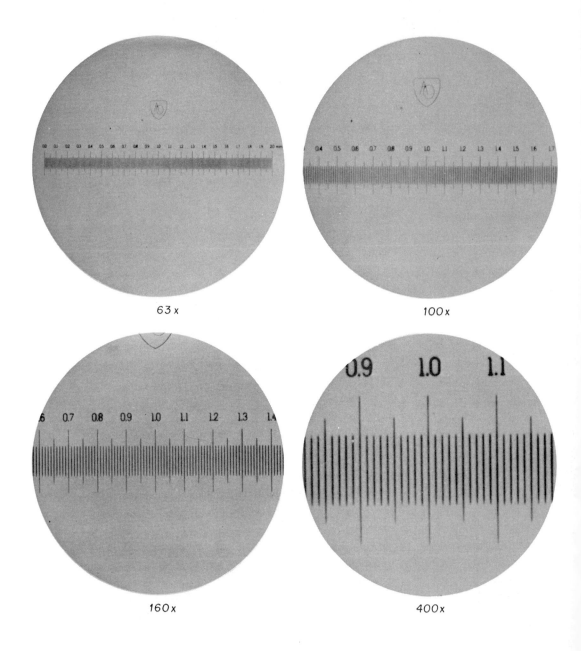

FIGURE 2-8. Field diameters. A given combination of ocular and objective on a given microscope produce a specific magnification, which is easily calculated, and specific field diameter, which can be measured using a stage micrometer.

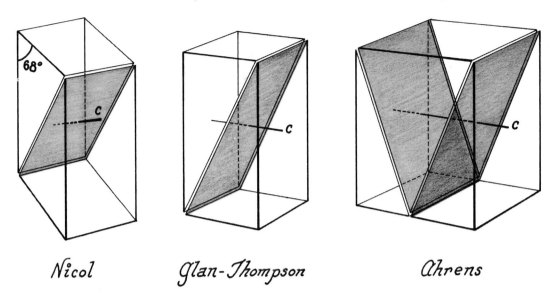

FIGURE 2-9. Polarizing calcite prisms. Prisms of three types are formed from optically clear calcite cemented by Canada balsam to produce a source of polarized light.

The Accessory Opening

Directly below the analyzer is an opening in the microscope tube through which the quartz wedge, the gypsum plate, and other accessories (discussed later) may be inserted into the light path (Figs. 2-2,*E* and 2-3,*E*). Some microscopes are equipped with devices for closing or plugging this opening when not in use to prevent dust from entering the microscope tube.

The Objective Lens

The critical component of any microscope is the objective lens (Figs. 2-2,*F* and 2-3,*F*), which acts with the ocular to accomplish the two basic functions of the microscope—magnification and resolution. The modern microscope objective is a complex system of lens elements that are formulated to compensate for defects inherent in lens systems. Because of the exceptionally low tolerance limits within which objectives are assembled, *they should never be disassembled by a novice.* The quality of a microscope objective is determined by the degree to which it corrects such defects as chromatic and spherical aberration (Fig. 2-10). Chromatic aberration is the result of failure of the

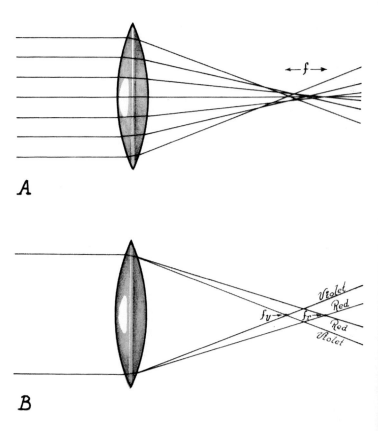

FIGURE 2-10. Lens aberration. (A) Spherical aberration results when rays passing near the periphery of the lens come to focus at a different point from those passing near the center. The center and edges of the visible field are not simultaneously in focus, and the field is said to be "dished." (B) Chromatic aberration results when light waves of different wavelength are brought to focus at different focal points and images show color fringes.

objective to focus light of different wavelengths at the same point. Spherical abberation is the result of failure of the lens to bring to focus at the same point the light that passes through its periphery and the light that passes through its center. The result is a "dished" field. An achromatic objective, or achromat, corrects chromatic aberration for two separate wavelengths, red and blue, and corrects spherical aberrations for one wavelength near the center of the visible spectrum. An apochromatic objective, or apochromat, is chromatically corrected for three wavelengths, red, yellow, and blue, and is spherically corrected for two separate wavelengths. Semi-apochromats, which contain an element of natural fluorite, provide a greater degree of correction than achromats but not as much as apochromats. Student microscopes are normally supplied with achromatic objectives, which are quite satisfactory for most petrographic work. Apochromats are useful for very critical work but require a special compensating ocular to complete their critical corrections. A third lens defect—one that is especially intolerable in polarizing microscopes—is mechanical strain, which makes optical glass "anisotropic" and produces anomalous double refraction. To eliminate disturbing internal reflections and to increase image brightness, both surfaces of all lens elements are coated with a thin film of an anti-reflection substance. Such substances have refractive indices substantially lower than that of optical glass.

Magnification is but one of two principle functions of the objective. Strain-free objectives for use with polarized light are available with practically any magnification, ranging from 1× to 100×. Most student-model petrographic microscopes are supplied with one low-power objective (e.g., 3.2×) for observing such large-scale features as rock texture, one medium-power objective (e.g., 10×) for general observation, and one high-power (e.g., 45×) for observing interference figures and small-scale features. The magnification specified on an objective lens is obtained only if the lens is used with a microscope having a tube length for which the lens was designed. The tube length, usually 160 mm or 170 mm, is the distance between the seating surface of the ocular and the objective flange.

Resolution, or resolving power, is the second principal function of an objective. Qualitatively, the resolving power of the objective is its ability to reveal fine detail. Quantitatively, it is a measure of the smallest distance at which two points on a magnified subject can be seen as clearly separate points. It is a function of the wavelength (λ) of the light used, the refractive index (n) of the medium between the objective and the subject, and the angle (u) between the central ray and the most highly refractive ray to enter the objective (Fig. 2-11,A):

$$\text{limit of resolution} = d = \frac{\lambda}{2n \sin u} = \frac{1}{\text{resolving power}}.$$

The quantity $n \sin u$ is called the numerical aperture (N.A.) and is directly proportional to resolving power. When air is the medium between subject and objective ($n = 1$), the N.A. cannot exceed 1.00. An obvious way to increase the N.A., and hence the resolving power, is to interpose between objective lens and subject a medium with a refractive index higher than that of air. This is precisely the principle behind the use of oil-immersion objectives

(Fig. 2-11,B). The immersion oil, which fills the space between lens and subject, has a refractive index approximately equal to that of the lens glass ($n = 1.515$).

It is of no value to use an immersion oil with a refractive index higher than that of the lens glass, for the index of the lens would become a limiting factor. Attaining maximum resolving power also requires an oil, or water, contact between subject slide and substage condensing lens. With oil immersion, a N.A. of about 1.5 can be attained, and with violet light ($\lambda = 4{,}000 \times 10^{-8}$ cm) the minimum limit of resolution for a light microscope proves to be about 1.3×10^{-5} cm.

$$\text{Min. limit of resolution} = \frac{\lambda}{2 \text{ N.A.}} = \frac{4{,}000 \times 10^{-8} \text{ cm}}{2 \times 1.5} = 1.3 \times 10^{-5} \text{ cm.}$$

Ultraviolet microscopy further decreases the limit of resolution by making use of wavelengths below the visible range.

A useful relation between the N.A. and the practical limit of useful magnification is

$$\text{max. useful magnification} \simeq \text{N.A.} \times 1{,}000.$$

Although high-power oculars can be combined with objectives to exceed their useful magnification, the extra enlargement is called "empty magnification." Since magnification and resolution are independent, resolution does not improve with an increase in magnification; when the useful magnification is exceeded each object point in the image becomes a small disc, called an Airy disc, and the resulting image is not only blurred and lacking in detail but may show unreal structures.

Because a cover glass acts as a lens element in the optical system, objectives are designed for use with a cover glass of specific thickness, usually 0.17 mm. Cover-glass thickness is not critical with low-power objectives, and for objectives with N.A. less than 0.3 the subject may even be viewed without a cover glass. For high-power objectives with N.A. greater than 0.7, a deviation as small as ± 0.01 mm from the optimum value of 0.17 mm seriously impairs the resulting image; even a layer of Canada balsam between subject and cover glass effectively thickens the cover glass. Consequently, specifically designed objectives are available that can be adapted by means of correction collars for use with cover glasses of specific thicknesses in the range 0.12 mm to 0.22 mm.

By agreement among manufacturers, objective lenses produced after about

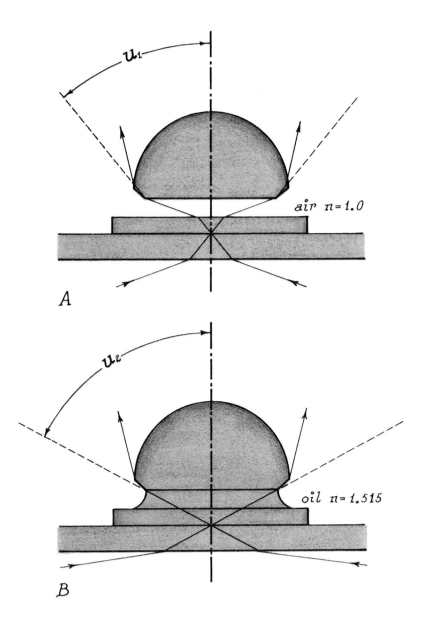

FIGURE 2-11. Oil immersion increases numerical aperture. Numerical aperture is proportional to the refractive index of the medium between the subject and the front lens of the objective and to the size of the cone of light that enters the front lens (i.e., N.A. = $n \sin u$). (A) In air the front lens of the objective captures but a small cone of light because much of the light is lost by refraction and by total reflection at the interface between air and cover glass. (B) In oil the front lens of the objective captures a larger cone of light because the oil and the lens have the same index of refraction and because no refraction occurs at the interface between oil and cover glass.

1950 are marked with magnification, numerical aperture, mechanical tube length, and cover-glass thickness (Fig. 2-12).

The free working distance of an objective is the distance between lens and cover glass when the subject is in clear focus. For low-power objectives it may be several centimeters, but for high-power objectives, it is a fraction of a millimeter, which means that extreme care should be taken to avoid scratching the lens by overfocusing it into the cover glass. Some high-power objectives are protected by a built-in spring mount that allows the front lens to retract into its barrel when overfocused into the cover glass. For those that are not so protected, it is safest to lower them close to the cover glass while observing from one side of the stage, and then focus by raising the objective. The extremely small free working distance of oil-immersion objectives may require the use of extra-thin cover glasses and unusual care in focusing.

Depth of focus is the vertical distance that is simultaneously in focus. Unfortunately, it decreases rapidly with increasing magnification and numerical aperture. As a result, high-power objectives are poor for general observations.

FIGURE 2-12. The objective lens. Markings on the objective lens denote its magnification, 40, its numerical aperture, 0.65, the tube length with which it was designed to be used, 160, and the cover-glass thickness at which it will give the best image, 0.17. The designation Pol indicates that the objective is strain-free and thus acceptable for use with polarized light. Other designations that sometimes appear on lenses indicate the type: Plan planachromat, Planapo planapochromat, Oel oil-immersion, and so forth. A lens without any such designation is a simple achromat.

Each petrographic microscope is designed in such a way that objectives may be changed simply and quickly. Most microscopes manufactured today have a multiple-lens nose turret that allows objectives to be interchanged simply by rotating the turret. Because some objectives are centered by rotating collars on the lens barrel, the lens barrels should not be used as handles for rotating the turret. Parfocal objectives are designed in such a way that the distance between object and virtual image is uniform for all objectives; thus when the turret is rotated to change objectives, only minor adjustments of focus need be made. In vertical-tube microscopes the objectives normally screw into a centering ring or collar held against the microscope tube by a spring-loaded objective clutch arm (Fig. 2-2,F_1). Once an objective is properly centered in a well-built microscope it can be removed and returned to the light path or be removed and replaced by another without the need for recentering.

Arm and Base

The frame of the microscope consists of an arm and a base (Figs. 2-2,G,H and 2-3,G,H). Both have the function of support. In vertical-tube microscopes the arm is normally connected to the base by an inclination joint that allows both tube and stage to be inclined. Although many such microscopes may be used conveniently when inclined, the practice is usually discouraged for petrographic work. It is preferred that the stage be in a horizontal position so that liquids will not run off and so that the slide can rest freely for ease of movement on the stage. Inclined-tube microscopes have no inclination joint; arm and base are built as a single frame.

The focusing adjustment is incorporated in the microscope arm. In straight-tube microscopes, a large knob on the upper part of the microscope arm is used for coarse focusing and a smaller one for fine focusing. The coarse adjustment rotates a small pinion that engages a rack on the back of the microscope tube; the mechanism for the fine adjustment may consist either of a small screw thread or a small cam and lever. Inclined-tube microscopes usually have a single knob on the lower part of the microscope arm with coaxial coarse and fine adjustments for lowering and raising the stage rather than adjusting the microscope tube. The fixed microscope tube better accommodates such accessories as cameras and vertical illuminators and does not require readjusting them each time the microscope is refocused. The fine adjustment is calibrated like a micrometer head for reading the vertical tube movement.

The Microscope Stage

The rotating stage (Figs. 2-2,*K* and 2-3,*K*) is marked off in degrees, and turns against a fixed vernier scale that can be read in tenths of a degree (European models) or in minutes (American models). Modern stages tend to be of large diameter for accuracy of measurement, and are mounted on a double row of ball bearings for smooth rotation under the heavy load of such accessories as the universal stage. A small knob at the rim of the stage (Figs. 2-2,K_1 and 2-3,K_1) allows the stage to be clamped in any position. Some stages have a second knob that activates a click-stop that works at 45° intervals. Most microscope stages have five holes in the upper surface; the two unthreaded holes are used for mounting spring clamps that hold glass sides in place, and the three threaded holes are used for mounting a mechanical stage. Stages that are built to accommodate a universal stage have two extra holes and a removable center plate.

The Mechanical Stage

Designed to permit a glass slide to be moved smoothly on the microscope stage in mutually perpendicular directions, the mechanical stage (Fig. 2-13) screws firmly to the microscope stage and rotates with it. A set of jaws for holding the slide can be moved in two mutually perpendicular directions by rack and pinion mechanisms controlled by small knobs. Movement in both directions is measured by vernier scales. Any area of interest on a particular slide can be recorded in terms of a pair of scale readings. Slides may thus be removed and returned at will, and any area of interest on the slide can be moved into the visible field by means of the verniers. Mechanical stages are also useful for counting procedures and other quantitative methods.

The Substage Apparatus

Beneath the rotating stage is the substage (Fig. 2-14), an assembly consisting of the polarizer, or lower nicol, an iris diaphram, and commonly one or more filter holders, all of which can normally be raised and lowered as a unit, either by a rack and pinion or by a screw mechanism (Figs. 2-2,*P* and 2-3,*P*). The substage unit can normally be swung out of the light path or be completely removed by lowering it to its minimum position after removing the substage mirror and its forked mount.

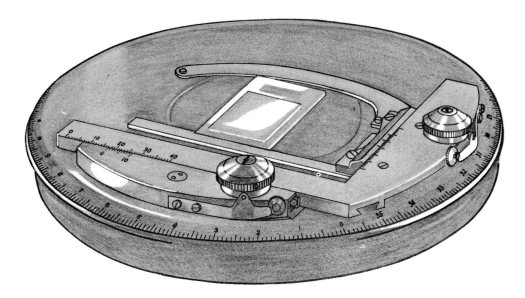

FIGURE 2-13. Mechanical stage. A mechanical stage facilitates systematic examination of a microscope slide by allowing measured movements in mutually perpendicular directions. It is used mostly for grain or point counting and other quantitative measurements.

THE CONDENSER (*Fig. 2-14,L*). In most student microscopes the condenser consists of two simple lens elements, but in research microscopes the condensing lens system may rival the objective in complexity. Its function is to supply a cone of light just large enough to fill the front lens of the objective completely, giving maximum illumination. The diameter of the light cone supplied by the condenser is a function of its numerical aperture (N.A.) and its distance below the subject. For best results the condenser should be held near its highest position, and its N.A. should be essentially equal that of the objective.

When calcite prisms form the polarizer, a comparatively narrow beam of light reaches the substage condenser because polarizing prisms have small cross sections. A permanent condensing lens (Fig. 2-14,*L*) with N.A. about equal to that of the medium-power objective (10×) is satisfactory for medium- and low-power objectives. A second condensing lens (Fig. 2-14,L_1) swings in above the permanent lens to condense the light cone further, to decrease the focal length of the condenser, and to increase its N.A. to essentially that of the high-power objective. When used with oil immersion objectives to attain maximum effective N.A., condensers with N.A. greater than 0.9 should have an oil, or at least water, contact with the bottom of the slide. A water contact between condenser and slide is usually satisfactory, and facilitates the work of cleaning and condenser system after use. The effective N.A. is one-half the sum of the N.A. for objective and condenser. When a large polaroid disk is used as polarizer, more light reaches the condenser; consequently, a single lens condenser is sometimes used with all objectives, although much light is wasted.

With the ocular removed the upper surface of the objective can be seen directly, and by raising or lowering the condenser the cone of light may be adjusted so that it just fills the objective.

THE IRIS DIAPHRAGM (*Fig. 2-14,M*). Under the condenser, usually just below its focal length, is the iris diaphragm. Its major function is to increase natural contrast through the formation of refraction halos (see p. 50), not merely to dim the image. Neutral gray filters should be used for controlling light intensity.

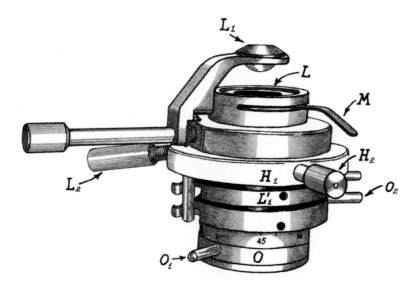

FIGURE 2-14. Substage assembly—Zeiss Standard GFL. The substage unit is an assembly of condensing lenses and the polarizer nicol. L_1. Swing-in, auxiliary condensing lens of large numerical aperture. M. Iris diaphragm. L. Permanent condensing lens of small numerical aperture below iris diaphragm. H_1. Substage support ring. H_2. Spring clamp for condenser assembly. Pull forward to release condenser assembly. L_2. Centering bolt for condenser assembly. L_1'. Auxiliary lens. O. Polarizer (lower nicol). Rotate with arm O_1, and swing out of field with arm O_2.

THE POLARIZER, OR LOWER NICOL (*Fig. 2-14,O*). The bottom element of the substage assembly is the polarizer, or lower nicol. The circular mount of the polarizer is held in place only by friction in most substages; consequently, caution should be used when the microscope is moved, as the polarizer is likely to fall unless it is properly seated. Most polarizers are disks of polaroid, but in some research microscopes a calcite prism of the Ahrens type may serve as a polarizer. Polaroid disks offer the advantage of requiring less vertical space, and they yield a broader cone of illumination. The polarizer can be rotated through 360° and can easily be slipped from its mount or swung from the light path for cleaning. The petrographic microscope is intended, however, to be used with the polarizer in place so that the visible field is always illuminated with polarized light. In most microscopes the polarizer is oriented so that it passes light waves vibrating only in a north-south plane when it is at its zero setting, but since the polarizer in some microscopes passes only waves that vibrate in an east-west plane, the student should be aware of the orientation of the polarizer in the microscope he is using.

THE SUBSTAGE MIRROR (*Fig. 2-2,Q*). Microscopes that do not have a built-in light source that directly illuminates the condenser are equipped with a circular substage mirror held in a forked mount below the substage assembly. The mirror reflects light from an illuminator into the optical system. One side of the mirror is flat and the other concave. The condenser is designed for use with the flat side of the mirror; the concave side produces astigmatism but may usefully enlarge the cone of light for low-power objectives. Most new microscopes have built-in light sources.

Accessories to the Petrographic Microscope

Many optical and mechanical accessories have been designed, some large and complex, others very simple. No attempt is made here, however, to consider all of them; some are so necessary to basic petrographic procedures as to be called "accessories" only because they are not physical parts of the microscope itself. The accessory box in most petrographic microscope cabinets contains, in addition to extra ocular and objective lenses, a quartz wedge, a gypsum plate, a mica plate, and commonly two wrenches or keys for centering the objectives. The quartz wedge, the gypsum plate, and the mica plate are merely described here. Their application will be taken up later.

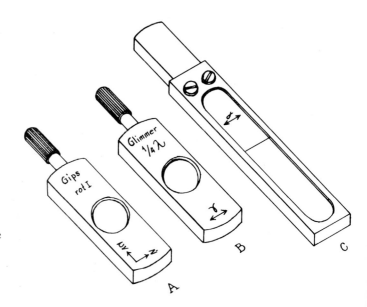

FIGURE 2-15. Microscope accessories. (A) The gypsum (1st-order red) plate. (B) The mica ($\frac{1}{4}\lambda$) plate. (C) The quartz wedge.

THE QUARTZ WEDGE (*Fig. 2-15,C*). This is indeed a wedge of the mineral quartz, measuring from zero thickness at one end to about 0.25 mm at the other. It is so cut from the quartz crystal that the *c*-axis of the crystal lies in the plane of the wedge perpendicular to the length of the wedge. This wedge is cemented to a piece of glass that is mounted in a metal holder made to fit the accessory opening of the microscope tube. An arrow at the thick end of the wedge is marked Z or γ, and indicates the vibration direction of the slow wave.

THE GYPSUM, OR FIRST-ORDER RED, PLATE (*Fig. 2-15,A*). This accessory is made either from a piece of selenite gypsum cut parallel to the perfect cleavage or from a piece of quartz cut parallel to the *c*-crystallographic axis. The plate, about 0.0625 mm thick, is cemented between two pieces of glass and mounted in a metal holder with the *c*-axis or Z optical direction (i.e., vibration direction of the slow wave) parallel to the arrow marked Z or γ on the mount (see p. 122). This accessory may be marked "1st-order red" or "retardation 550 mμ." Those manufactured in Germany are marked "Gips rot I."

THE MICA, OR $\frac{1}{4}\lambda$, PLATE (*Fig. 2-15,B*). This accessory consists of a cleavage sheet of optically clear muscovite mica about 0.03 mm thick placed between two glass discs and mounted in a metal frame. The Z optical direction is parallel to the arrow marked Z or N' on the metal frame. The mica plate is the quarter-wavelength accessory, and is marked "$\frac{1}{4}\lambda$." Those made in Germany are marked "Glimmer $\frac{1}{4}\lambda$."

Illuminators

Since daylight is extremely variable and not always available, artificial illuminators are used with blue daylight filters to remove yellow from tungsten illumination. Two basic types of illuminators are in common use. The simple type—the one generally encountered by students—consists of a tungsten filament bulb behind ground glass or opal glass, both of which provide uniform luminescence. The more complex illuminators supply an intense light from a small six- or eight-volt bulb controlled by a variable transformer. These units are generally equipped with a focusing condensing lens, an iris diaphram, and appropriate daylight and neutral filters. High-intensity, low-voltage illuminators are difficult to use because the filament source is small and supplies uneven illumination. Consequently, a ground-glass diffuser below the polarizer is usually used with these illuminators. For optimum results, light intensity is controlled by neutral filters. The lamp iris supplies a beam of light just large enough to illuminate the visible field, and the substage iris and condenser form a cone of light just large enough to fill the objective.

Proper Use of the Microscope

Proper use implies proper care of the instrument and maximum efficiency and comfort for the user. The student should realize that the microscope and accessories entrusted to his care represent an investment of perhaps $1,000 and are delicate, precision scientific instruments. The microscope and all accessories should be placed in the cabinet provided or be properly covered with a dustproof covering when not in use. The microscope should be carried by its arm with the free hand below its lower nicol. Dust may be removed from painted metal surfaces by wiping with a soft cloth or tissue, but dust on the lens or prism surfaces should be removed only with a camel's hair brush; oil or fingerprints should be removed only by wiping them with lens paper. If a liquid is necessary, only distilled water (just breathe on the lens), ether, or xylene should be used. Do not use alcohol or acetone, as they may dissolve the cement between the lens elements. Optical glass is softer than common glass and demands exceptional care.

Dust on ocular lenses appears as discrete specks that rotate as the ocular is turned, oil on the objective lens causes a foggy field, and dust on the substage condenser may be seen with the ocular removed.

High-power lenses should be lowered almost to the cover glass and brought into focus by raising the microscope tube or lowering the stage.

The sliding surfaces of rack and pinion movements should occasionally be wiped clean and very lightly lubricated.

The observer should be able to see through the microscope from a normal sitting position. To avoid eye strain one should learn to keep both eyes open and concentrate on the microscope eye, alternating eyes occasionally. Eye glasses are unnecessary except for astigmatism.

Adjustment of the Microscope

Proper Illumination

To adjust illumination it is necessary to remove the ocular, the Bertrand lens, and the analyzer from the optical path so that the upper element of the objective is visible from the upper end of the microscope tube. The condenser is raised or lowered until its limiting diaphragm fills the diaphragm of the objective, the mirror is tipped so that maximum illumination is centered in the objective diaphragm, and the illuminator is then moved away from or toward the mirror until the objective lens is filled with even illumination. Brightness is controlled by neutral filters on the illuminator. The swing-out condenser is used with objectives, above about 25×. To achieve so-called critical illumination (also called "Köhler illumination"), the illuminator is placed about eight inches from the substage mirror with its iris diaphragm nearly closed, and the subject is brought into focus. The image of the diaphragm is then formed in the plane of the subject by raising or lowering the substage condenser until the diaphragm image is sharp, and the lamp diaphragm is opened until the visible field is just fully illuminated. The ocular is now removed and the substage iris is opened or closed until the upper element of the objective is fully illuminated. The ocular is then replaced and the microscope is ready for use.

Centering the Objective Lens

The objective must be centered in its mount so that the central point of the visible field remains centered as the stage is rotated (Fig. 2-16). The objective

FIGURE 2-16. Centering the objective lens. (A) Stage rotation causes conspicuous grains to rotate about a center far removed from the center of the visible field. (B) Adjusting centering screws or rings moves the center of the field (i.e., intersection of cross hairs) toward the center of rotation. (C) The objective is centered when center of rotation and center of field coincide.

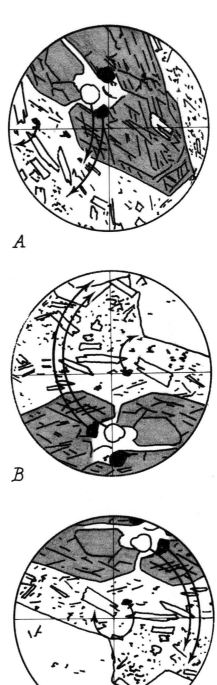

is moved in its collar or barrel by turning small bolts on the collar or by rotating rings on the objective barrel. Rotating the stage causes the entire field to rotate about some fixed point in or near the visible field. If this fixed point lies at the intersection of the cross hairs, the objective is centered; if not, the fixed point is moved to the intersection by turning centering bolts or collar rings.

Some microscopes have fixed objectives and movable stages. These are centered by means of two small bolts that move the stage against a spring mount. Inherent in this design is a certain amount of instability of the stage plate. In a few specially designed research microscopes, like the Leitz "SY," or the B & L "LD," the problem of objective centering is completely avoided by eliminating the need for stage rotation. Analyzer and polarizer nicols can be rotated simultaneously, so that the sample remains undisturbed. This is an especially useful advantage when some bulky apparatus, such as a hot stage is being used on the microscope stage. (A hot stage is an apparatus designed to heat a sample while it is being viewed.)

Proper Alignment of the Polarizer and Analyzer Nicols

The polarizer normally passes light waves that vibrate in a north-south plane, and unless the polarizer has been loosened in its mount it is properly aligned when the 0°-position of the mount is aligned with its reference mark. This alignment can be quickly and easily checked by using some mineral that is known to absorb light vibrating in a readily recognizable crystallographic direction. For example, the *c*-crystallographic axis of a tourmaline crystal is parallel to its length, and the crystal absorbs more light vibrating perpendicular to its length than it does light vibrating parallel to it. As the stage is rotated, a tourmaline crystal lying flat on the microscope stage will, through uncrossed nicols, appear darkest when its *c*-axis is perpendicular to the plane of the polarized light (i.e., if the crystal shows maximum color when oriented east-west, the polarizer is properly aligned). Similarly, biotite strongly absorbs light vibrating parallel to its cleavage and passes light vibrating perpendicular to it. A flake of biotite on a microscope slide will, of course, show no cleavage, but biotite is common in many rocks whose thin sections contain crystals of biotite with cleavage more or less perpendicular to the plane of the section. Such a crystal should appear darkest when the cleavage is north-south (Fig. 2-17).

With the lower nicol properly aligned, the vibration direction of the upper nicol is easily determined. Since the anlayzer should pass only light vibrating

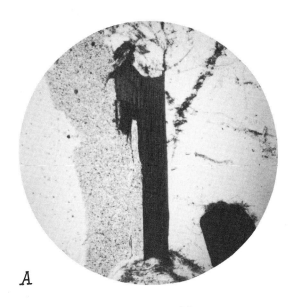

A

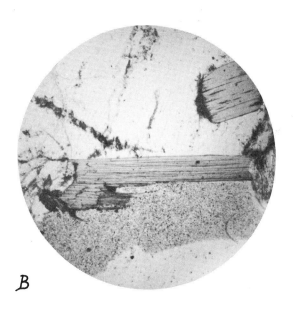

B

FIGURE 2-17. Absorption of light by biotite. (A) Biotite strongly absorbs light waves vibrating parallel to its cleavage. The analyzer is passing N–S waves. (B) Biotite weakly absorbs light waves vibrating perpendicular to its cleavage. The analyzer is passing N–S waves.

in an east-west plane, it should pass no light from the lower nicol. If, therefore, the field is dark when the analyzer is placed in the light path, the nicols are properly aligned and said to be *crossed*. If the field is not black, the upper nicol is not properly oriented. Most student microscopes do not allow manual manipulation of the upper nicol, which should be adjusted only by an expert.

CHAPTER 3

Refractometry

General Principles

The index of refraction of a transparent substance may be found by measuring either the critical angle or the angles of incidence and refraction (see pp. 11–12). Several methods for measuring these angles are commonly used for liquids and homogeneous solids.

The only practical method of measuring the index of refraction of microscopic particles was first described by O. Maschke in 1872; it consists in comparing the index of a particle with that of a surrounding medium. A colorless, transparent medium, like water or window glass, is visible only because the light rays are refracted and reflected at the interface between it and the medium that surrounds it. If a colorless substance and its surrounding medium have exactly the same refractive index the substance is invisible, since light rays pass from one medium into the other without deviation. It is precisely this phenomenon that makes it possible to measure the refractive index of mineral particles—by immersing them in a series of liquids of known indices until a liquid is chosen whose index matches that of the particles. For this purpose, elaborate sets of immersion liquids of known refractive index are available.

Relief

The degree of visibility of a transparent particle in an immersion medium is called its relief (Fig. 3-1). A fragment having the same index of refraction as the liquid in which it is immersed is invisible and is said to have zero relief; a fragment whose refractive index is considerably higher or lower than the liquid stands out in high relief—positive relief if its index is higher than that of the liquid, and negative relief if lower. Relief is expressed qualitatively by such terms as "low," "moderate," "high," or "extreme," but it may also be expressed quantitatively as the numerical difference between the refractive indices of particle and immersion medium. Since Canada balsam or some synthetic resin of similar refractive index (i.e., about 1.537) is used as the mounting medium for rock thin sections, this index is normally used as the standard for comparison. In anisotropic substances the index of refraction differs with direction. A fragment showing high relief in one orientation and low relief in another is said to have variable relief.

Dispersion

Dispersion is the separation of polychromatic light into its component wavelengths. It becomes apparent when the different wavelengths are diffracted at different angles. Since the component wavelengths have different velocities, the refractive index of a given medium will differ for each wavelength. The quantitative measure of dispersion is a pure number equal to the difference between the index of refraction of the transmitting medium for one end of the visible spectrum and the index for the other. It is commonly represented as

$$\text{Dispersion} = n_F - n_C$$

or

$$\text{Dispersion} = V = \frac{n_D - 1}{n_F - n_C},$$

where n_F, n_D, and n_C are the refractive indices of the transmitting medium for the $F(\lambda = 486$ mμ; blue-green), $D(\lambda = 589$ mμ; yellow), and $C(\lambda = 656$ mμ; red-orange) wavelengths (Fraunhofer lines) of the solar spectrum, respectively.

In white light, an immersed mineral fragment seldom shows zero relief because of the difference in the dispersion produced by the fragment and that produced by the immersion medium. Unless their dispersions are identical, they must show relief for some wavelengths of the visible spectrum. Immersion

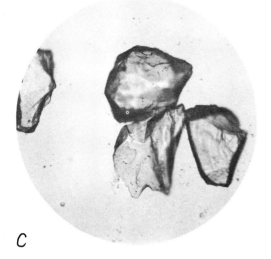
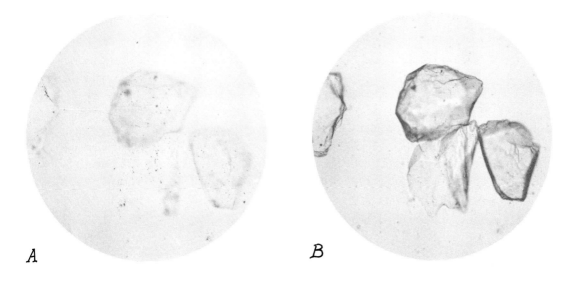

FIGURE 3-1. Relief is the degree of visibility of fragments or crystal sections in their surrounding media. The quantitative value of relief is the numerical difference in refractive index. (A) Low relief. Quartz fragments ($n_\omega = 1.544$) in immersion oil with $n = 1.550$. (B) Moderate relief. Quartz fragments ($n_\omega = 1.544$) in immersion oil with $n = 1.576$. (C) High relief. Quartz fragments ($n_\omega = 1.544$) in immersion oil with $n = 1.74$.

oils usually show higher dispersion than immersed fragments. When fragment and oil have the same refractive index for wavelengths at the middle of the spectrum, the index of the fragment is usually less than that of the oil for shorter wavelengths and more than that of the oil for longer wavelengths (Fig. 3-6), and colored dispersion fringes result. If indices are the same at wavelengths near the middle of the spectrum, color fringes appear pale blue on one side of the interface and pale yellow on the other. If indices are the same at long wavelengths (i.e., near red), the fringe on one side of the interface will contain all wavelengths except red and appear almost white, and the fringe on the other side will contain only long wavelengths and appear distinctly red or orange. If indices are the same at short wavelengths, the fringes are nearly white and distinctly violet or blue.

Immersion oils in the range below 1.60 usually have low dispersion, as do most minerals, with the result that color fringes are weak. High-index oils, however, show extreme dispersion, and color fringes may spread to a complete spectrum.

The index of refraction that a given crystal shows for red light may be as much as 0.06 higher than it shows for violet; consequently, we may expect to measure its refractive index in white light only to an accuracy of ± 0.03. The only way to eliminate dispersion, and thereby increase accuracy of measurement, is to use a monochromatic light. A sodium vapor lamp ($Na_D \lambda = 589$ mμ) is commonly used to measure refractive index. When indices of refraction are quoted for minerals or immersion oils, they are generally understood to be for this wavelength unless stated otherwise. The use of monochromatic light should improve the accuracy of refractive-index measurements by at least one order of magnitude. If sources of monochromatic red light (C) and monochromatic blue light (F) are available,* the numerical dispersion of a mineral can be determined by measuring the index of refraction for those wavelengths. The refractive index of an immersion oil also changes with wavelength and must be determined for the wavelengths used.

Central Illumination, or Becke Line, Method

Perhaps the most widely used method of comparing indices of particle and immersion medium is that of central illumination, originally described by F. Becke in 1893. To improve sensitivity, natural contrast is increased by reducing the aperture of the substage iris diaphragm. Sensitivity can also be improved somewhat by increasing magnification, but a 10× objective is satisfactory for most index measurements.

Mineral fragments vary widely in shape but are nearly always thicker in the center than near their edges. Thus they act as crude lenses, either concen-

---

*Colored glass filters, available from Eastman Kodak Co., Corning Glass Works, and other manufacturers, produce "essentially monochromatic" red, blue, and even "sodium" light from normal tungsten radiation. A monochromator is, of course, a better source of monochromatic light but is seldom available.

trating or dispersing the light that passes through them. If a fragment has a higher index than the immersion medium, it acts as a crude converging lens, concentrating light within the fragment image. When the focus is raised, light rays tend to converge as a "bright blur" at the center of the fragment image. When the focus is lowered, light rays separate, and the brightness moves outward. If the fragment has an index lower than the immersion medium, it becomes a "diverging lens," and the light rays diverge from it as the focus is raised. This effect becomes more difficult to observe as relief becomes greater. A fragment that shows extreme positive relief may concentrate light rays so that some of them cross before entering the objective. As the focus is raised, these rays diverge, and others may converge: the entire effect becomes confused. If the initial immersion medium has the same index as Canada balsam, high relief usually indicates positive relief, since few minerals have refractive indices much lower than that of balsam.

If the relief is low, the light rays only slightly converge or diverge and appear as a thin bright line called the *Becke line* at the interface between fragment and immersion medium (Fig. 3-2). If the fragment has the higher index, the Becke line lies on the fragment side of the interface and moves inward as the focus is raised; if the immersion medium has the higher index, the line moves outward. *As the microscope tube is raised, or the stage lowered, the Becke line moves toward the medium of higher index.*

Minerals in thin section and some mineral fragments may be of uniform thickness and therefore will not act as lenses. Even if the mineral-liquid interface is vertical, however, a Becke line will form by concentration of light rays in the higher index medium by refraction and total reflection at the interface (Fig. 3-3).

When the refractive index of the immersion medium is almost equal to that of the fragment, dispersion may produce a red and blue Becke line; and, in thin section, colorless minerals of low negative relief often appear faintly pink, and those of low positive relief faintly blue, to the experienced eye.

Oblique Illumination Method

Simple and widely used, the oblique illumination method of index comparison is based on the use of directional illumination. To produce such illumination, some elaborate microscopes have a decentering iris diaphragm below the substage condenser which can be moved to darken half the field of view. Since most microscopes are not so equipped, it is necessary to block the light from half of the optical path by inserting a card above the condenser or below the polarizer, by tipping the mirror or covering half of it, or by inserting an acces-

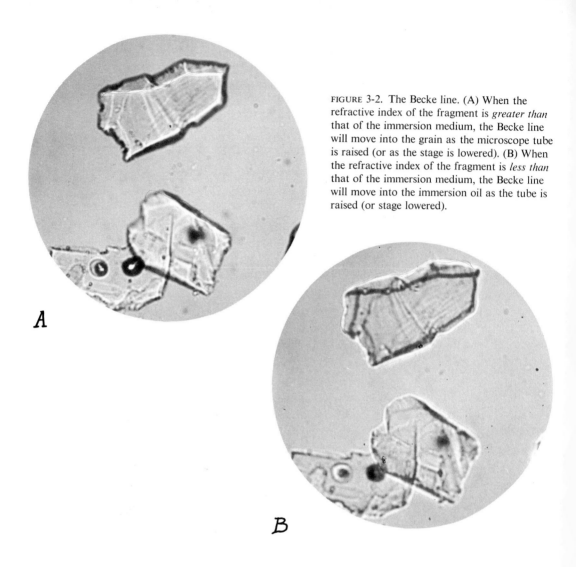

FIGURE 3-2. The Becke line. (A) When the refractive index of the fragment is *greater than* that of the immersion medium, the Becke line will move into the grain as the microscope tube is raised (or as the stage is lowered). (B) When the refractive index of the fragment is *less than* that of the immersion medium, the Becke line will move into the immersion oil as the tube is raised (or stage lowered).

sory halfway into the microscope tube. Fragments appear bright on one side and dark on the other when so illuminated (Fig. 3-4). This phenomenon, too, is the result of the convergence or divergence of light by refraction at the interface, as shown diagrammatically in Figure 3-5. From the figure, it is apparent that if the fragment has the higher index, it should be dark on the side toward the illuminated half of the field. But the optics of the microscope sometimes reverse this relationship, which depends on the position of the card blocking the illumination, the position of the substage condenser, and the combination of lenses used. Since the method can cause confusion, it is generally advisable for the student to examine a fragment of known index to determine which side is illuminated under a given set of conditions; if these conditions are religiously duplicated thereafter, confusion may be avoided.

This method does have several advantages over central illumination: it may be used to advantage with low-power magnification and is reportedly somewhat more accurate for very critical investigations.

High Dispersion Method

An immersion liquid that shows high dispersion embraces a nearly continuous range of refractive indices within its limits of dispersion—a property that may be used to advantage in combination with the central illumination or, Becke line, method described earlier.

When the dispersion of an immersed fragment is the same as the dispersion of the immersion liquid, the Becke line is a distinct white line, and the refractive index of the fragment is either greater or less than that of the liquid for all wavelengths. For refractive indices less than 1.60, dispersion of both liquid and fragment is usually very small. Dispersion tends to increase with refractive index at a faster rate for the liquid, with the result that at high refractive indices, the dispersions of liquid and fragment are significantly different.

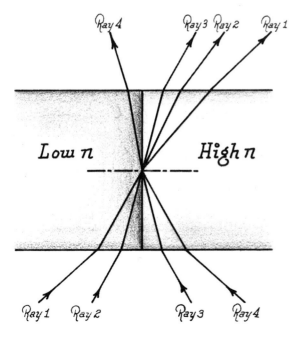

FIGURE 3-3. Formation of the Becke line. The Becke line is formed near the interface between media of high and low refractive index by the concentration of refracted light rays that converge at the interface. At the interface, pairs of equally convergent rays (1 and 4 and 2 and 3) are refracted toward the interface normal in the medium of high index and away from the normal in the medium of low index. Note that rays which strike the interface at incidence angles greater than the critical angle (e.g., ray 3) are totally reflected, producing a concentration of rays (Becke line) in the medium of high refractive index.

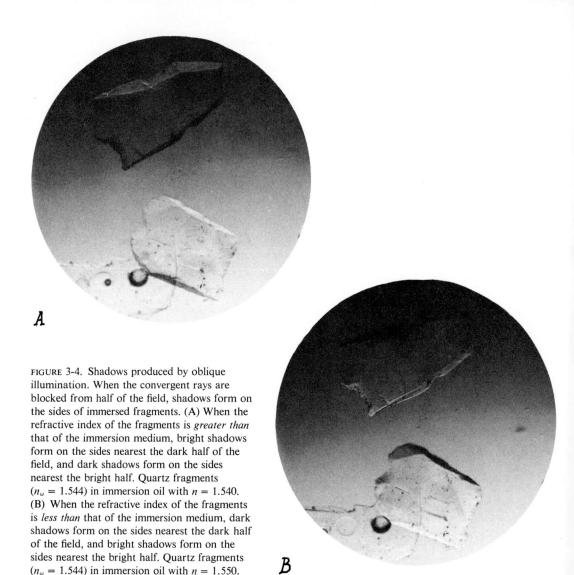

FIGURE 3-4. Shadows produced by oblique illumination. When the convergent rays are blocked from half of the field, shadows form on the sides of immersed fragments. (A) When the refractive index of the fragments is *greater than* that of the immersion medium, bright shadows form on the sides nearest the dark half of the field, and dark shadows form on the sides nearest the bright half. Quartz fragments ($n_\omega = 1.544$) in immersion oil with $n = 1.540$. (B) When the refractive index of the fragments is *less than* that of the immersion medium, dark shadows form on the sides nearest the dark half of the field, and bright shadows form on the sides nearest the bright half. Quartz fragments ($n_\omega = 1.544$) in immersion oil with $n = 1.550$.

Thus the refractive index of fragment and liquid can be the same at only one wavelength (Fig. 3-6). The Becke line may no longer be a distinct white line on one side of the fragment-liquid interface, but may consist of two colored bands, one made up of short wavelengths and the other of long wavelengths. If the difference in dispersion between fragment and liquid is small, the spread of the Becke line will also be small, but if the difference in dispersion is great, the Becke line will appear as a complete spectrum. If the Becke line does

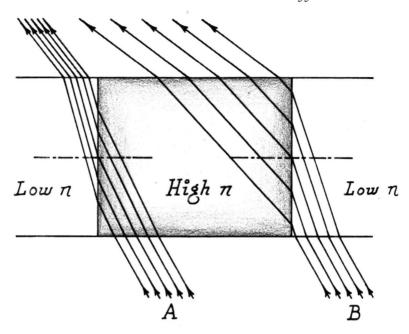

FIGURE 3-5. Formation of the shadows of oblique illumination. Oblique light rays, coming from one side only, are refracted toward the interface normal when entering a medium of higher refractive index, and away from the interface normal when entering a medium of lower refractive index. When the immersed fragment possesses a higher refractive index than its immersion medium, refraction produces a concentration of light rays (i.e., bright shadow) on the side nearest the dark half of the field and a separation of light rays (i.e., dark shadow) on the side nearest the illuminated half of the field.

appear as a spectrum, one may be able to observe, while the focus is raised and lowered, which spectral color remains immobile at the fragment-liquid interface as the other colors move in and out across the interface. This color is produced by the approximate wavelength at which fragment and liquid have identical refractive indices. If we know the dispersion of each of two immersion liquids and can find the wavelengths at which the refractive index of the fragment agrees with that of each liquid, we can establish a line relating refractive index of the fragment and wavelength, and n_D for the fragment lies on the line at $\lambda = 589$ mμ (Fig. 3-7).

Practical difficulties arise in judging the wavelength (i.e., color) at which the indices of fragment and liquid agree, and accurate judgment may require considerable experience. The greater the dispersion produced by the liquid, the broader the Becke line spectrum and the more accurate the judgment. For low values of liquid dispersion, the Becke line spectrum is narrow and appears divided into pale color fringes; one can usually judge only the broad range of wavelengths within which agreement occurs. Although the accuracy of the method depends on the accuracy with which the wavelength of index agreement can be determined, one may expect to measure the refractive index of an unknown fragment to an accuracy of about ± 0.002.

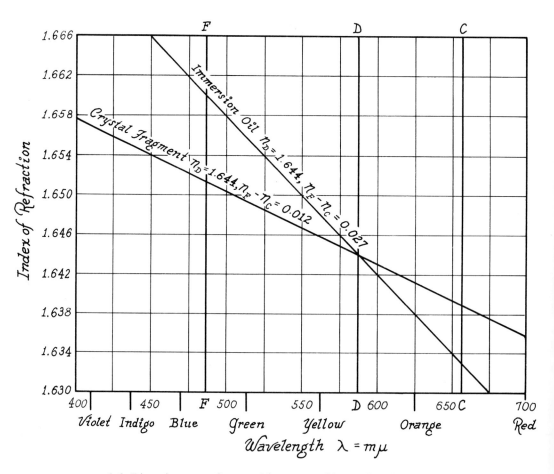

FIGURE 3-6. Dispersion curves for crystal fragment and immersion oil. Dispersion curves, assumed to be linear, may be plotted for immersion oil and crystal fragment if the refractive index for wavelength D ($\lambda = 589$ mμ) and the dispersion are known for each. For each curve, n_D establishes a point and the dispersion ($n_F - n_C$) establishes the slope. If the curves have different slopes, fragment and oil can have the same refractive index for only one wavelength.

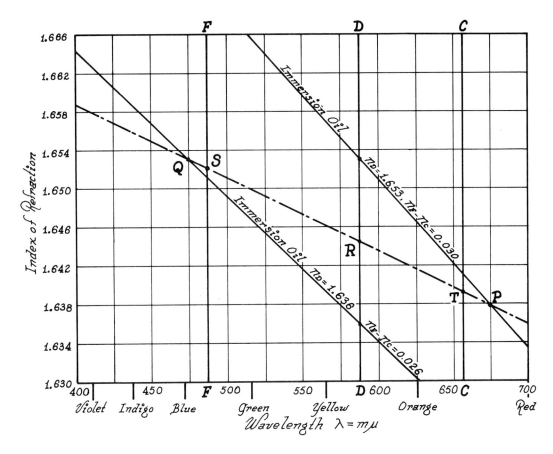

FIGURE 3-7. Finding the refractive index of a crystal fragment by using two high-dispersion immersion oils. The refractive index of a crystal fragment has been found to be the same as that of an immersion oil with $n_D = 1.636$ for a blue wavelength (Q) and the same as that of an immersion oil with $n_D = 1.653$ for a red-orange wavelength (P). Line PQ is the dispersion curve of the crystal fragment. The value of n_D for the fragment is found at R; the dispersion of the fragment is obtained from the values of n at S and T ($n_F = n_C = 1.652 - 1.639 = 0.13$).

Other Methods

Although the central and oblique illumination methods are the most commonly used, several other procedures are employed when greater accuracy of measurement is required. Most require elaborate apparatus and are generally beyond the scope of introductory courses. The following discussion gives a brief outline of the underlying principles.

The "double variation" method of R. C. Emmons is so called because it is based on the use of two independent variations in measurable properties.

The refractive index of the immersion medium can be varied by changing its temperature. Increasing the temperature decreases the refractive index of the immersion liquid while that of the immersed fragment remains virtually unchanged. Each immersion liquid can therefore be made to vary continuously over a range of indices. The other variation used in this procedure is that of the wavelength of monochromatic light. Since immersion liquids generally show greater dispersion than solids, the index of the liquid varies faster with change of wavelength than does that of the immersed solid. By varying temperature and wavelength separately or together, it is possible to measure the index of a solid quite accurately for any given wavelength. The method has obvious advantages for experienced workers, but the special warming stage with its thermometer and temperature control, the temperature-controlled refractometer in series with the stage and the monochromator used to vary wavelengths are obvious disadvantages to the inexperienced investigator.

A double diaphragm of C. P. Saylor makes possible a simple refinement of the oblique illumination procedure; the device has one shading diaphragm below the condenser lens and another above the objective lens.

Accuracy of Refractive-index Measurements

The accuracy of refractive-index measurements is largely dependent on the experience of the investigator, but other considerations limit accuracy to about ± 0.0005 even under ideal conditions.

The refractive index of an immersed fragment can, of course, be measured to no greater accuracy than that of the index of its immersion medium. Immersion liquids calibrated to an accuracy of ± 0.0002 are prepared commercially, with refractive index recorded at a given temperature. A measured temperature coefficient (i.e., the change in index per degree of temperature change) makes possible a correction to the actual working temperature. If the room temperature is lower than that for which the index of the liquid was recorded, a correction obtained by multiplying the temperature coefficient (usually about 0.0004 per C°) by the temperature difference is added to the given refractive index.

For accurate measurements, it is desirable to have immersion liquids available in steps of 0.002 or even 0.001, but seldom do beginning students have access to such a set. To obtain such accuracy, it is therefore necessary to mix available liquids to get one that approximates the index of a fragment. The accuracy obtained depends on the difference in the indices of the two liquids, the care with which the relative volumes are measured, and the degree of linearity of the volume-index relationship. A good procedure consists in mixing the oils in a watch glass until the index matches that of the fragments and then measuring the refractive index of the oil with a refractometer.

Dispersion of the immersion medium is also a source of inaccuracy. High-dispersion liquids show a wide range of index for different wavelengths, and introduce more uncertainty than low-dispersion liquids. Since monochromatic light eliminates this source of error, regardless of dispersion, it is always used for critical determinations. *Note*: Contamination of immersion liquids by carelessness (e.g., interchanging bottle lids or dropper rods) will render measurements inaccurate.

Size, shape, color, and other variables affect the accuracy with which the index of refraction of mineral particles can be measured; these variables are difficult to consider quantitatively. Refractive indices of crystalline substances and noncrystalline substances that show only one index of refraction (i.e., isotropic substances) can be measured with greater accuracy than those of anisotropic crystals, whose indices vary with the direction of observation.

In the study of mineral optics, the most significant variable of all is completely beyond the control of the investigator. This is the variability of index within the same mineral species. Minerals with simple and fairly definite chemical formulas (e.g., fluorite, anhydrite, and quartz) are consistent to the third or fourth decimal place. Some minerals, however, frequently have indices that vary in the first or second decimal place, because they contain impurities, allow extensive ionic substitution, or have no definite chemical formula at all. For these reasons it is seldom worth trying to measure indices as accurately as possible; in fact, the beginning student will seldom find it necessary to measure mineral indices to an accuracy greater than ± 0.01.

Immersion Liquids

Characteristics of Immersion Liquids

Beginning with O. Maschke in 1872, many investigators have suggested various liquids for use with the immersion procedure. The ideal immersion liquid is colorless, odorless, nontoxic, and chemically inert; it has low volatility, low dispersion, low viscosity, and a low temperature coefficient; it retains its index of refraction for long periods of time with normal exposure to air and light; and it is completely miscible with similar liquids of higher and lower index, thus providing a continuous range of indices through proper blending.

The modern student is seldom concerned with the preparation of immersion media, since they are now commercially available with indices ranging from

TABLE 3-1. Properties of Immersion Media

Immersion media	Index of Refraction n_D at 20°C	Temperature Coefficient $-\dfrac{dn}{dt}$	Dispersion $(n_F - n_C)$	Specific Gravity	Remarks
Air	1.00027				
Water	1.333			1.00	Dissolves many low-index minerals
Acetone	1.357		Slight	0.79	Dissolves some minerals (volatile)
Ethyl alcohol	1.362	0.00040	Slight	0.79	Dissolves some minerals (volatile)
Ethyl butyrate	1.381		Slight	0.88	
Methyl butyrate	1.386		Slight	0.90	
Ethyl valerate	1.393		Slight	0.87	Very volatile
Heptane	1.400			0.68	Dissolves plastic bottle caps
Paraldehyde	1.406			0.99	
Amyl alcohol	1.409	0.00042	Slight	0.80	Dissolves some minerals
Kerosene	1.448	0.00035	Slight		
Cineole	1.458			0.92	
Petroleum oil					
Russian alboline	1.470	0.0004	0.012		
American alboline	1.477	0.0004	0.012		
Amyl phthalate	1.488			1.19	
Ethyl salicylate	1.523		0.021	1.14	Pleasant odor
Clove oil	1.531	0.00050	Moderate		Immiscible in petroleum oil
Methyl salicylate	1.536		0.015	1.18	"Oil of wintergreen"
Ethylene bromide	1.538			1.52	
Nitrobenzene	1.553			1.20	
Bromobenzene	1.560			1.50	
Benzyl benzoate	1.569			1.11	
Cinnamon oil	1.585–1.600	0.00030	Strong		Pleasant odor
Iodobenzene	1.621			1.83	
α-Monochloronaphthalene	1.625		Moderate	1.19	"Hallowax"
α-Monobromonaphthalene	1.658	0.00048	0.028	1.49	
Methylene iodide	1.738	0.00070	0.037	3.33	Metallic Sn or Cu prevents discoloration
Sulfur (10%) in arsenic tribromide	1.814	0.0007	Strong		Toxic and corrosive

TABLE 3-1, continued

Phenyldiiodoarsine	1.838		0.40	Toxic
Sulfur (20%) and arsenic disulfide in arsenic tribromide	2.003	0.0006	Strong	Toxic and corrosive
Arsenic tribromide, arsenic disulfide and selenium	2.11		Very Strong	Toxic and corrosive
Sulfur and selenium	1.998 2.716		Very Strong	Low temperature melts
Sulfur and arsenic selenide	2.72 3.17		Very Strong	Low temperature melts

1.35 to 2.11 in steps of 0.002 for the low and intermediate range and in steps of 0.01 for the high-index liquids.* The indices of most minerals lie well within this range, but to cover the mineral index range completely it would be necessary to provide liquids to beyond 4.0.

The low- and intermediate-index liquids are generally a mixture of those suggested by Larson and Berman (see Table 3-1). Liquids above 1.74 are generally patterned after two high-index series described by Meyrowitz and Larsen in 1951 and 1952. One series is a mixture of bromonaphthalene and a solution of precipitated sulfur (10 percent) in arsenic tribromide, and ranges from 1.66 to 1.81. Another series, ranging from 1.81 to 2.00, is a mixture of precipitated sulfur (10 percent) in arsenic tribromide and precipitated sulfur (20 percent) and arsenic disulfide in arsenic tribromide (60 percent). These liquids vary in color from pale to deep amber, are extremely toxic and corrosive, and should not come in contact with metal or lens surfaces. They readily attack human tissue and should be removed from skin immediately.

Low-melting compounds of sulfur, selenium, and arsenic may be used to extend the index range to 3.17. The solid is slowly melted on a glass slide, the mineral sprinkled into the melt, a cover glass pressed down on the melt, and the melt allowed to cool. Heating too long or at too high a temperature will cause changes in refractive index. Most melts are highly colored and require monochromatic red (lithium) light for all examinations. Melts of TlCl, TlBr, and TlI show indices between 2.4 and 2.8, and are considerably more transparent.

---

*R. P. Cargille Laboratories, Inc., 117 Liberty St., New York 6, N.Y.

USE OF IMMERSION LIQUIDS. These liquids are kept in small bottles with a glass dropper rod connected to a ground-glass stopper or screw cap. One way to prepare a mount is to apply a drop of immersion liquid directly to the glass microscope slide by touching it with the rod, a few mineral fragments are then added to the drop, and a cover glass is placed over the drop with a pair of forceps (to avoid fingerprints). Some prefer to place fragments on the slide first, before applying the liquid. Others prefer to place the fragments on the slide and then cover them before adding the immersion liquid; a drop of liquid applied to the edge of the cover glass will be drawn in by capillary action. If it proves necessary to mix liquids, this should be done with a separate glass rod. Mixing can be done right on the slide or in a watch glass. When small cover glasses are used, two mounts can often be made on one standard 26 mm × 45 mm microscope slide. It is preferable to begin with an intermediate liquid (e.g., 1.54), and it is good practice to mark the index of the liquid on the slide or lay the slide on a piece of paper denoting the index so that mounts can be compared later.

Most immersion liquids are soluble in xylene, which is used extensively for cleaning slides and cover glasses. A final washing with water and detergent removes the oily film. High-index liquids containing dissolved sulfur and arsenic are not soluble in xylene; consequently, it is usually advisable to discard the slide. These liquids are, however, soluble in toluene or carbon bisulfide.

Gelatin-coated slides are manufactured by the Eastman Kodak Co. A drop of water on the slide causes the gelatin to swell, and fragments sprinkled on this surface are held fast when the gelatine dries. Successive immersion liquids can be applied to the same fragments and washed away with xylene without disturbing the orientation of the fragments. A mechanical stage facilitates return to a particular field after changing liquids.

Measuring the Refractive Index of Liquids

THE REFRACTOMETER. Although the student seldom finds it necessary to make up a set of index liquids, he may need to verify the refractive index of liquids supplied him. The quickest, simplest, and most conventional method of measuring the refractive index of a liquid is by refractometer, an instrument designed expressly for this purpose. The sturdy and convenient Abbe refractometer consists basically of a telescope and two refracting prisms (Fig. 3-8) separated by a layer of the liquid whose index is to be determined. As the telescope is rotated, the refracting prisms also rotate until the light path exceeds the critical angle at the interface between prism and liquid and the field of the telescope is half light and half dark. The arc-shaped scale, which moves

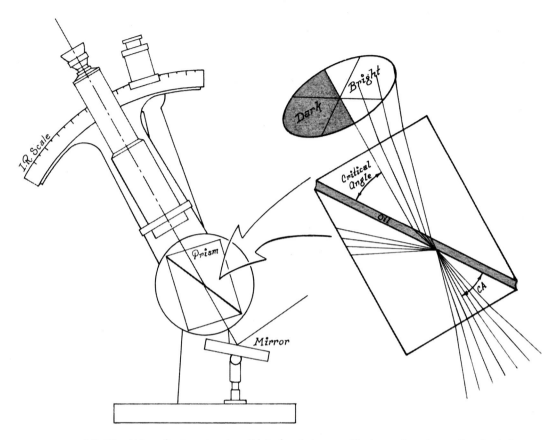

FIGURE 3-8. The Abbe refractometer. An oil interface between split glass prisms transmits refracted light rays at all angles of incidence less than the critical angle, which is measured on an arc scale that can be read directly in index of refraction.

with the telescope, is calibrated directly in refractive index, which is read directly to the third decimal place and estimated to the fourth. Since an immersion liquid shows dispersion (greater at high indices), its refractive index differs for red and violet, and a spectrum separates the light and dark halves of the field. This spectrum becomes a sharp line when monochromatic light (usually sodium 589 Å) is used. Two Amici prisms in the telescope tube rotate simultaneously in opposite directions to minimize dispersion.

Because the refractive index of the refracting prisms must exceed the index of the liquid being measured, the Abbe refractometer may be used only for liquids whose indices are between 1.3 and 1.7. Immersion liquids with indices higher than 1.74 often contain constituents that may chemically attack or mechanically scratch the soft glass of the prisms.

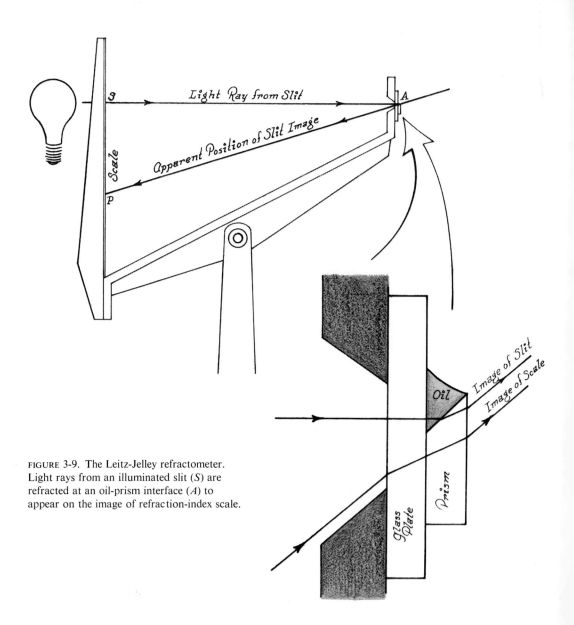

FIGURE 3-9. The Leitz-Jelley refractometer. Light rays from an illuminated slit (S) are refracted at an oil-prism interface (A) to appear on the image of refraction-index scale.

THE LEITZ-JELLEY MICRO-REFRACTOMETER. The standard model of this simple refractometer affords a range of n_D from 1.333 (water) to 1.92, with measurements corrected to ± 0.001. A special prism extends the range to 2.350.

The refractometer consists basically of a vertical scale with a light slit at 1.52 and a micro-prism (refractive index = 1.52) mounted behind a small opening that is in line with the slit (Fig. 3-9). A tiny drop of immersion oil is placed in the groove between the prism and the glass plate on which it is mounted, and one looks through the small opening to see two images. One image is the scale, seen unrefracted through the glass plate, and the other is the illuminated slit after refraction at the oil-prism interface. Where the slit image appears on the scale, the index of refraction is read directly. If immersion oil and prism have the same refractive index (i.e., 1.52), the ray forming the slit image is not refracted at the oil-prism interface, and the image appears at its true position at 1.52. If the refractive index of the oil is less than 1.52, the image is refracted downward at the prism; if greater, it is refracted upward.

If sodium light is used to illuminate the slit, its image is always a narrow band, but if white light illuminates the slit, its image may appear as a complete spectrum, in which case n_D is measured at the yellow band. A filter may be inserted between the light source and slit to yield a distinct dark band in the spectrum at the yellow position. The width of this spectral image depends upon the amount of dispersion produced by prism and oil. Low-dispersion oils have about the same dispersion as the prism, and produce a sharp white slit image, but high-dispersion oils produce a broad spectrum.

Minimum Deviation Method

A very accurate way of measuring the index of a solid is to form a prism of that substance and measure its angle of minimum deviation (Dm).

To measure the refractive index of a liquid, a hollow glass prism is filled with the liquid. (The surfaces of the prism walls are parallel so that they produce no deviation.) The angle of minimum deviation and prism angle obtained are those of the liquid. Several authors* have suggested simple methods for making such hollow prisms.

Consider a ray of monochromatic light (RR') pasing through a prism of angle A (Fig. 3-10). By passing through the prism, the ray is refracted from its original course by angle D (i.e., the angle of deviation). By constructing a line parallel to oo', it can be seen that

$$D = (\theta + \theta') = (i - r) + (i' - r') = (i + i') - (r + r').$$

By constructing a second line parallel to oo', it becomes clear that

$$A' = r + r',$$

---

*E. S. Larsen and Harry Berman, Bull. 848, U.S.G.S. (1934), pp. 18–20.

and, from perpendicular sides, we know that

$$A' = A = r + r' \quad \text{and} \quad D = i + i' - A.$$

if n is the refractive index of the prism,

$$n = \frac{\sin i}{\sin r} \quad \text{and} \quad n = \frac{\sin i'}{\sin r'},$$

$$\sin i = n \sin r \quad \text{and} \quad i = \sin^{-1}(n \sin r),$$

$$\sin i' = n \sin r' \quad \text{and} \quad i' = \sin^{-1}(n \sin r') = \sin^{-1}[n \sin(A - r)].$$

Therefore, since $D = i + i' - A$,

$$D = \sin^{-1}(n \sin r) + \sin^{-1}[n \sin(A - r)] - A.$$

The angle of deviation (D) reaches a minimum when dD/dr equals zero.

$$\frac{dD}{dr} = \frac{n \cos r}{\sqrt{1 - n^2 \sin^2 r}} - \frac{n \cos(A - r)}{\sqrt{1 - n^2 \sin^2(A - r)}}.$$

If $dD/dr = 0$,

$$\frac{n \cos r}{\sqrt{1 - n^2 \sin^2 r}} = \frac{n \cos(A - r)}{\sqrt{1 - n^2 \sin^2(A - r)}}.$$

Squaring both sides and substituting $r' = A - r$, we get

$$\frac{n^2 \cos^2 r}{1 - n^2 \sin^2 r} = \frac{n^2 \cos^2 r'}{1 - n^2 \sin^2 r'}.$$

It follows that $r = r'$ when the angle of deviation is the minimum angle of deviation (Dm).

If $r = r'$, then

$$i = i' \quad \text{and} \quad a = a',$$

$$Dm = a + a' = 2a = 2(i - r) = 2i - 2r,$$

$$i = \frac{Dm + 2r}{2},$$

But $r = A/2$ and $i = (Dm + A)/2$. Since $n = \sin i / \sin r$ we have

$$n = \frac{\sin \tfrac{1}{2}(Dm + A)}{\sin \tfrac{1}{2} A}.$$

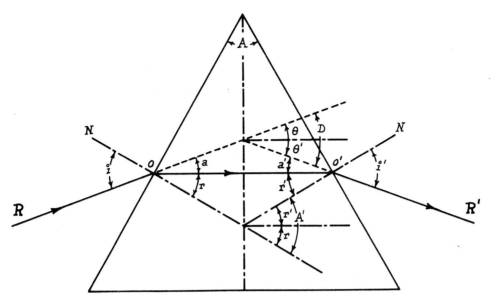

FIGURE 3-10. The angle of minimum deviation. The index of refraction of a prism is a function of prism angle A and angle of minimum deviation Dm:

$$n = \frac{\sin \frac{1}{2}(Dm + A)}{\sin \frac{1}{2} A}.$$

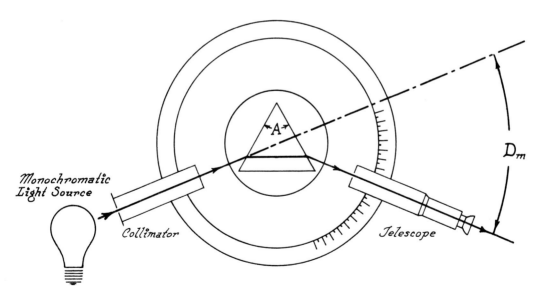

FIGURE 3-11. The index of refraction of a liquid can be measured by the method of minimum deviation with a hollow glass prism on a simple spectrometer.

The refractive index of a prism is, therefore, a simple function of the angle of minimum deviation and the prism angle.

In practice, the angle of minimum deviation (Dm) is measured with a simple spectrometer (Fig. 3-11). The telescope and prism are rotated until the minimum deviation is experimentally discovered, and the prism angle (A) is measured by using the spectrometer as a reflecting goniometer. Light is reflected from one prism surface, as from a mirror, and the prism is rotated until the second prism surface occupies the same position, as indicated by its reflection. The prism angle is 180° minus the angle of rotation.

CHAPTER 4

Isotropism and Isotropic Media

The Nature of Isotropism

The term "isotropic"—*iso*, meaning equal, and *tropic*, meaning to turn, rotate, or mix—is used to describe a medium that shows equal or uniform physical properties in all directions. In an isotropic medium, light is propagated with equal velocity in all directions because the atoms or molecules that make up the medium have random arrangement or equivalent arrangements along all crystallographic axes.

Gases and Liquids

A gas is composed of atoms or molecules that are free to move and are not bound together. Being very small compared to the volume in which they are free to move, they are essentially independent of each other, neglecting occasional collisions. Although the individual molecules of some gases are strongly polarized (i.e., not symmetrical) their random arrangement renders them statistically uniform, and thus isotropic. Similarly, most liquids are composed of nonsymmetrical molecules in statistically random arrangements, and are therefore isotropic.*

---

*Under special conditions the asymmetrical molecules of some liquids may assume a systematic orientation and form so-called "liquid crystals," which are not isotropic.

Amorphous Solids

Amorphous solids are essentially supercooled liquids whose atoms or molecules have no systematic repetition and are therefore isotropic. Since glass, air, and liquids give us our major acquaintance with transparent media, one has little need for the concept of isotropism unless he undertakes a study of crystalline solids, which of course is what we are about to do.

Isometric Crystals

Many of our most familiar crystalline compounds are also isotropic because they crystallize in the isometric crystal system. Although the atoms that compose such crystals are systematically arranged, their arrangement is identical along three mutually perpendicular crystallographic directions (Fig. 4-1). Should such a crystal become strained, this would no longer be true and the crystal would be anisotropic. Many isometric minerals, such as halite, sylvite, and fluorite, form ionic crystals in which the outer electron shell of every ion consists of eight electrons symmetrically distributed about the nucleus. Thus when light passes through such a medium, it is equally absorbed and refracted regardless of its direction of propagation. A few isotropic minerals (e.g., diamond) are covalent, with a symmetrical bond distribution.

Behavior of Light in Isotropic Media

Since an isotropic medium is one which transmits light with uniform velocity in all directions, it follows that the refractive index of the medium is the same for all directions of transmission and that the vibration directions of polarized or nonpolarized light waves are not restricted on entering or leaving such a medium.

Imagine light rays radiating in all directions from a point source within an isotropic medium. The ray-velocity surface (i.e., the surface formed by the ends of all advancing rays at any given instant of time) is a sphere whose radius increases in direct proportion to the time of transmission and to the velocity of light within the medium (i.e., in inverse proportion to refractive index).

Isotropic Minerals and the Petrographic Microscope

Plane polarized light from the lower nicol is unchanged by its passage through an isotropic fragment on the microscope stage, and is thus eliminated

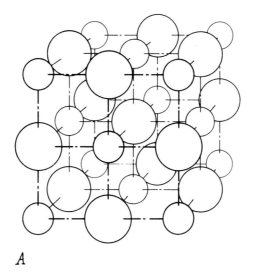

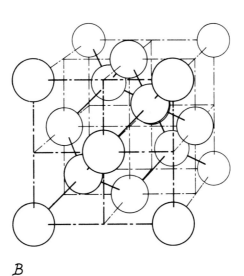

FIGURE 4-1. Isotropic crystal structure. (A) The crystal structure of halite (NaCl) is representative of ionic structures and shows alternating Na^+ and Cl^- ions on all three mutually perpendicular crystallographic directions. (B) The only mineral with a completely covalent structure is diamond. Each carbon atom is bound to four other identical atoms in tetragonal distribution by equivalent sp^3 bonds to form a three-dimensional isometric network.

by the upper nicol. Isometric and amorphous mineral fragments are therefore *dark* between crossed nicols, and herein lies the simplest test for anisotropism. (Although a given fragment of an anisotropic mineral may be dark between crossed nicols because of special orientation, all isotropic minerals are dark regardless of orientation.) The few aids to identification of isotropic minerals with the petrographic microscope are discussed below.

Index of Refraction

The only diagnostic optical property of an isotropic mineral that can be measured with a petrographic microscope is its single index of refraction (n).

Color

Color is sometimes helpful in identifying isotropic minerals. Selective light absorption must be uniform in all directions, and isotropic mineral fragments are never pleochroic (see p. 115). This does not preclude uneven color distribution in minerals like fluorite.

Cleavages and Structures

Cleavage cannot exist in amorphous minerals and, if present in isometric minerals, is cubic, octahedral, or, rarely, dodecahedral (Fig. 4-2). Although amorphous minerals commonly display banded colloform structures, such structures as accicular, reticulate, fibrous, radiated, and foliated are never shown by amorphous minerals, nor are they found in crystals with isometric symmetry.

Hardness and Specific Gravity

Hardness and specific gravity may sometimes be crudely measured with a petrographic microscope. By rubbing two glass slides together with a few mineral fragments between, one may determine the relative hardness of mineral and glass.

One should not overlook the implications of mineral fragments observed to float in the immersion oil beneath the cover glass. The specific gravities of many common minerals are less than those of some common immersion oils,

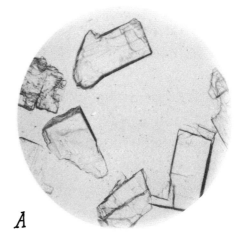

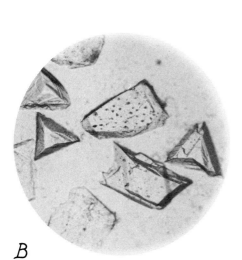

FIGURE 4-2. Isometric cleavage. Isometric crystals may show little or no cleavage but those that do display cubic, octahedral, or dodecahedral cleavage. (A) Isometric crystals with cubic {100} cleavage have three equivalent, mutually perpendicular cleavages. (B) Isometric crystals with octahedral cleavage {111} have four equivalent cleavages, often yielding triangular shapes. (C) Isometric crystals with dodecahedral cleavage {110} have six equivalent cleavages.

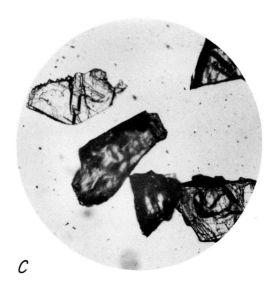

especially oils of high refractive index. Specific gravities of a few common immersion liquids are listed in Table 3-1. Although the student may not always know the specific gravity of the immersion oil he is using, he may be quite sure that the oil of index 1.74 is methylene iodide (diiodomethane), which has a specific gravity of 3.3.

CHAPTER 5

Optical Crystallography of Uniaxial Crystals

Uniaxial Anisotropism

An anisotropic medium is one in which light is transmitted with different velocities in different directions.

Anisotropism and Crystal Structure

A polarized light ray is one in which the wave vibrations are not symmetrical about the ray, and a polarized atom is one in which the orbiting electrons are not symmetrical about the nucleus. A single isolated atom contains orbiting electrons in statistically uniform distribution about the nucleus, but when this atom is acted upon by the electrical fields of surrounding atoms or the electrical vibrations of passing light waves, the density of its orbiting electrons is greatest in the direction of greatest electrical attraction and the atom is said to be polarized. The degree of polarization depends upon how strongly orbiting electrons are held by the nucleus, and only outer, loosely-held electrons are strongly polarized by external forces. The polarization of atoms in an isotropic medium is either statistically random (amorphous substances) or directionally uniform (isometric crystals), and the atoms are not consistently polarized by their atom neighbors. In anisotropic crystalline media, however, atoms experience nonuniform electrical attraction from dissimilar neighboring atoms to become polarized in the direction of strongest attraction. All equivalent atoms in the structure are similarly polarized by equivalent environments. The degree of polarization of an atom depends on the element and its bonding relations to surrounding atoms; large anions (distant orbiting electrons), strongly directional bonds, and low-order atomic coordination favor strong polarization. The plane of maximum polarization is usually the plane of maximum atom population density, and the plane of minimum polarization is usually perpendicular to it. The atoms of chain or sheet structures are most strongly polarized in the plane of the sheet or direction of the chain.

A light wave is an oscillating electrical field superimposed upon existing polarization forces; the wave may act either to increase or decrease the atomic polarization caused by structural environment. The polarization of an atom strongly affects the velocity of light waves transmitted by that atom. In general, the stronger the polarization of an atom, the slower the velocity (i.e., the higher the index of refraction) of the light wave that acts to increase the polarization. Optical anisotropism results when the ionic or covalent polarizing forces are unevenly distributed in the crystalline structure, with the result that transmitted light waves encounter different polarization environments in different crystallographic directions. Since only isometric crystals have symmetrically distributed bonding forces, *all substances that crystallize in the tetragonal, hexagonal, trigonal, orthorhombic, monoclinic, and triclinic crystal systems are anisotropic.*

The crystal structure of calcite ($CaCO_3$), shown in Figure 5-1, is analogous to the simple structure of halite (NaCl), in which Ca^{++} and CO_3^{--} ions replace the alternate Na^+ and Cl^- ions. The carbonate ions are arranged perpendicular to one cube diagonal, which becomes the unique or c-crystallographic axis, and the cubic cell becomes a rhomb. The carbonate ion consists of one small carbon atom surrounded by three oxygen atoms that form the corners of an equilateral triangle. The ion is held tightly together by strong, largely covalent bonds that are much stronger than any external bond and are highly polarized in the plane of the ion. Light waves transmitted parallel to the c-crystallographic axis vibrate parallel to the planes of carbonate ions, increasing polarization already present. Calcite therefore shows a relatively high index of refraction and a correspondingly low velocity for waves advancing parallel to c, regardless of their vibration direction normal to the axis. Light waves passing through calcite perpendicular to the c-crystallographic axis advance with different velocities depending upon their vibration directions. Waves vibrating perpendicular to c also vibrate in the plane of the

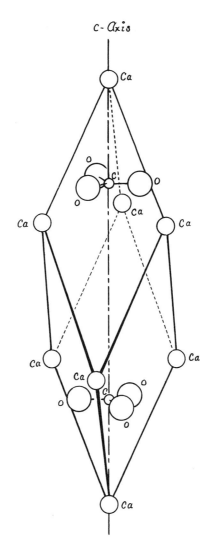

FIGURE 5-1. The structure of calcite.

carbonate ions, increasing polarization and advancing with the same velocity as waves moving parallel to c. Waves vibrating parallel to c vibrate at right angles to the plane of the carbonate ions, thus reducing polarization and advancing with greater velocity (i.e., the index of refraction is lower).

The "Strange Refraction of Calcite"

The behavior of light in calcite can be observed by viewing a dot on a piece of paper beneath a calcite rhomb. Two images, separated both horizontally and vertically, are seen within the rhomb. Light waves originating at the dot are refracted along two separate paths by the calcite crystal. The two images thus produced are responsible for the term "*double refraction*," the quantitative measure of which is called *birefringence*.

The everyday experience of viewing objects submerged in water, where the image of an object appears shallower than the object itself, is a good example of what is stated in the law of Duc de Chaulnes: the image depth (d_i) is inversely proportional to the index of refraction of the transmitting medium, and directly proportional to the true depth (d). (See Fig. 5-2.)

$$n = \frac{\sin i}{\sin r} = \frac{\tan i}{\tan r} = \frac{\frac{ab}{d_i}}{\frac{ab}{d}} = \frac{d}{d_i} \quad \text{or} \quad d_i = \frac{d}{n}.$$

The two images in the calcite rhomb lie at different depths because the two rays that form them experience different indices of refraction and advance with different velocities; the ray that forms the deeper of the two images encounters the lowest refractive index and is thus the faster ray.

If the calcite rhomb is rotated about a vertical axis, the deep image is observed to move in a circle about the shallow image, which remains stationary. When observed normal to the rhomb surface, the light ray that produces the shallow, stationary image is not refracted, in accord with Snell's law, and is designated the *ordinary ray*. The ray that forms the deep, moving image must be refracted, in defiance of Snell's law, and thus is called the *extraordinary ray*.

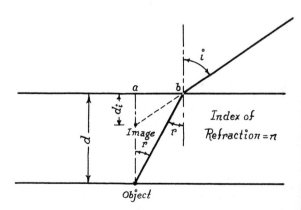

FIGURE 5-2. The Law of Duc de Chaulnes. Image depth and true depth are related by the expression $d_i = d/n$.

A polaroid sheet laid over the calcite rhomb can be used to demonstrate that light waves moving along ordinary and extraordinary rays are plane polarized in mutually perpendicular directions. When the polaroid is oriented to pass waves vibrating perpendicular to the c-crystallographic axis of the calcite rhomb, only the shallow, stationary (i.e., ordinary) image appears. When the polaroid is rotated 90°, only the deep (i.e., extraordinary) image can be seen.

A sphere cut from a calcite rhomb will yield even more experimental facts. When the sphere is viewed normal to the c-crystallographic axis, two dot images appear at maximum vertical separation but with no horizontal separation. Rotation of the sphere about the c-axis produces no change in the image positions, but rotation about a horizontal axis (i.e., normal to c) causes the images to separate horizontally and to approach one another vertically; moreover, while the ordinary image remains stationary, the extraordinary image moves horizontally away from and then back toward the stationary dot and rises to merge with the ordinary image when the line of observation merges with the c-direction. Observation along the c-axis yields only one dot at the position of the ever-stationary, ordinary image. (The index of refraction is constant in the c direction.) The polaroid sheet may be used to show that light rays moving parallel to c are not polarized.

Uniaxial Ray Velocity Surfaces

Derivation of Uniaxial Ray Velocity Surfaces

Observation of refraction in calcite leads to the following conclusions:

1. Ordinary rays behave as though they were being transmitted by an isotropic medium, obeying Snell's law and moving with equal velocity in all directions.

2. All light waves moving parallel to the c-crystallographic axis move with the velocity of ordinary rays and are not polarized.

3. Extraordinary light rays do not behave as though they were being transmitted by an isotropic medium; they defy Snell's law of refraction and move with different velocities in different directions.

4. Along c, the extraordinary ray becomes ordinary, but away from c its velocity increases over that of the ordinary ray and reaches a maximum when its direction of propagation is perpendicular to c.

5. Ordinary waves always vibrate normal to their rays, and normal to the c-crystallographic direction.

6. Extraordinary waves always vibrate normal to the vibration direction of the ordinary waves in a plane parallel to the c-direction.*

A ray velocity surface is the surface formed by the ends of all light rays emanating from a point source. We regard it as a physical phenomenon that is real and could be observed if our techniques of observation were sufficiently keen. Imagine a point light source embedded within a large calcite crystal. Suppose that at a given instant, light rays commence to advance outward from this point in all directions, and that shortly thereafter we halt the advance of all light rays and examine the surface formed by the ends of these rays. Instead of the simple sphere of an isotropic medium, there would be two separate geometrical surfaces, one formed by ordinary rays and one by extraordinary rays. Since ordinary rays travel with equal velocity in all directions, *the ray velocity surface formed by all ordinary rays must be a sphere* centered about the point source whose radius is proportional to the velocity (V_ω) or inversely proportional to the index of refraction for ordinary rays (n_ω).† *All waves of ordinary rays vibrate normal to the rays and normal to the c-crystallographic direction.*

The ray velocity surface formed by extraordinary rays is an oblate spheroid of revolution***tangent on its minor axis (i.e., c-axis) to the enclosed sphere (Fig. 5-3,B).* The minor, or rotation, axis of the spheroid is proportional to the ordinary ray velocity (V_ω), and the major axis is proportional to the maximum velocity (V_ϵ) or inversely proportional to the minimum refractive index (n_ϵ) experienced by the extraordinary rays. *All waves from extraordinary rays vibrate normal to the vibration direction of the corresponding ordinary wave in a plane parallel to the c-crystallographic direction.*

---

*Although the vibration of the ordinary wave is fixed by the above description, that of the extraordinary wave is not completely fixed, since its vibration direction may lie anywhere within a plane parallel to c. As a first approximation, the vibration direction of an extraordinary wave may be considered to be normal to its corresponding ray, but it is of academic interest to note that unless the ray is moving parallel or perpendicular to c, this is certainly no more than a first approximation. A complete explanation of this problem requires many pages of text and an understanding of a few principles that are difficult for the beginning student and of little practical value to him. The problem will therefore be ignored.

†Symbols n_ω and n_ϵ used herein are equivalent to ω, or n_O, and ϵ, or n_E, used by certain authors.

**An oblate spheroid of revolution is the surface formed by an ellipse rotated about its minor axis, and a prolate spheroid of revolution is the surface formed by an ellipse rotated about its major axis.

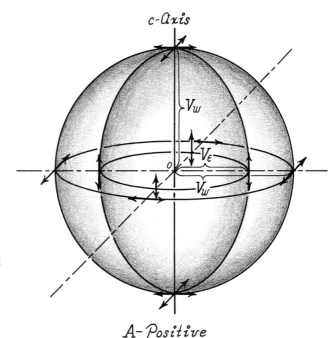

FIGURE 5-3. Uniaxial ray-velocity surfaces. Light rays within any uniaxial crystal advancing outward from a point of origin O produces two concentric surfaces. One surface is a sphere produced by ordinary rays that advance in all directions with identical velocities (V_ω). The second surface is a spheroid of revolution, either prolate (positive) or oblate (negative), formed by extraordinary rays that advance with different velocities in different directions. The crystal is defined as optically positive if ordinary rays are faster than extraordinary ones (A) and negative when the extraordinary rays are faster (B).

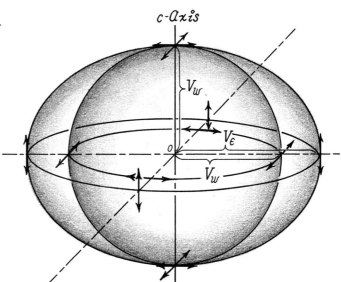

Had we selected quartz as the transmitting medium, rather than calcite, the ray velocity surfaces would consist of one prolate spheroid of revolution enclosed within a sphere, and the surfaces would be tangent at the c-crystallographic axis (Fig. 5-3,A). Again the sphere is formed by ordinary rays of velocity V_ω and the prolate spheroid by extraordinary rays moving at maximum velocity (V_ω) along c and at minimum velocity (V_ϵ) normal to c.

The Optic Sign

Ray velocity surfaces for quartz show that the ordinary rays advance more swiftly than the extraordinary rays in all directions except parallel to the unique c-crystallographic direction. *Uniaxial crystals, like quartz, in which the ordinary ray is faster (i.e., $V_\omega > V_\epsilon$ and $n_\omega < n_\epsilon$) are optically positive. Conversely, uniaxial crystals, like calcite, in which the ordinary ray is slower (i.e., $V_\omega < V_\epsilon$ and $n_\omega > n_\epsilon$) are optically negative.*

The Optic Axis

Ray velocity surfaces of any uniaxial mineral, positive or negative, have only two points in common, and these define a line parallel to the unique crystal direction. *Such a line gives the direction* of the optic axis, parallel to which all light rays advance with uniform velocity; the corresponding waves are free to vibrate in any direction normal to the light path. Crystals showing only one unique crystallographic direction (i.e., tetragonal, trigonal, and hexagonal) have only one optic axis (always parallel to the c-crystallographi axis), and are designated uniaxial crystals.* Biaxial crystals have two directions of uniform light transmission; in an isotropic medium any direction is an optic axis.

The Uniaxial Indicatrix

Construction of the Uniaxial Indicatrix

In 1891 Fletcher proposed an indicatrix, a purely imaginary geometrical surface constructed to represent, or "indicate," vibration directions and indices of refraction of light waves in an anisotropic medium (Fig. 5-4). The indicatrix of a *uniaxial positive* crystal is a *prolate spheroid of revolution* with the length

---

*Note that an optic axis is a direction and not a single line.

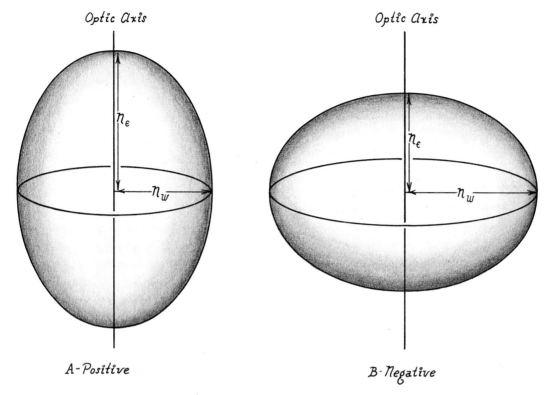

FIGURE 5-4. The uniaxial indicatrix. An indicatrix is a spheroid of revolution, either prolate (A) or oblate (B) constructed to indicate the vibration directions and refractive indices of light waves advancing in any direction within a uniaxial crystal. An indicatrix is constructed so that half of its axis of rotation is n_ϵ, representing the maximum (positive) or minimum (negative) index of refraction of all extraordinary rays, and its radius of rotation is n_ω, representing the refractive index of all ordinary rays. The properly constructed indicatrix is oriented with its optic axis, or axis of rotation, parallel to the c-crystallographic axis of the crystal it represents. For light rays moving normal to the optic axis or parallel to it, a central indicatrix section normal to the light ray yields an ellipse or circle that represents vibration directions (i.e., the directions of the axes of the ellipse) and refraction indices of waves associated with each vibration direction (i.e., the lengths of the semi-axes of the ellipse).

of the semi-major axis (i.e., axis of revolution) equal to the maximum refractive index of the crystal for extraordinary rays (n_ϵ) and the semi-minor axis (i.e., radius of revolution) equal to the refractive index for ordinary rays (n_ω). The indicatrix of a *uniaxial negative* crystal is an *oblate spheroid of revolution* with the length of the semi-minor axis (i.e., axis of revolution) equal to the minimum refractive index of the crystal for extraordinary rays (n_ϵ) and the semi-major axis (i.e., radius of revolution) equal to the refractive index for the

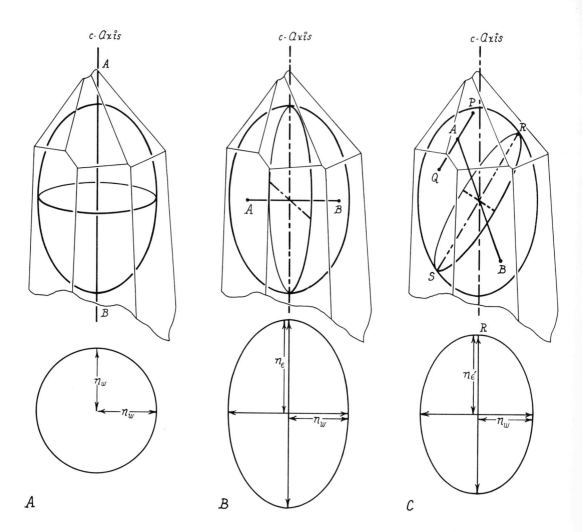

FIGURE 5-5. How to use an indicatrix. A quartz crystal (uniaxial, positive) is shown with its imaginary indicatrix appropriately constructed (i.e., $n_\omega = 1.544$ and $n_\epsilon = 1.553$) and properly oriented with the direction of its optic axis parallel to the c-crystallographic axis.
(A) Light-propagation direction AB is parallel to the optic-axis direction. A section normal to AB, through the center of the indicatrix, is a circular section of radius n_ω, as shown. *Conclusion:* Light waves moving along AB are not restricted to vibrate in any given direction (i.e., waves may vibrate parallel to any diameter of the circular section) and experience an index of refraction of n_ω. (B) Light-propagation direction AB is normal to the optic-axis direction. A section perpendicular to AB through the center of the indicatrix is an ellipse of maximum ellipticity. The length of the semi-major axis equals n_ϵ, and that of the semi-minor axis equals n_ω. *Conclusion:* Light waves moving along AB are confined to vibrate parallel to the major and minor axes of the elliptical section. The ordinary wave vibrates parallel to the minor axis (i.e., perpendicular to the optic axis) and experiences a refractive index of n_ω. The extraordinary wave vibrates parallel to the major axis (i.e., parallel to the optic axis) and experiences a refractive index of n_ϵ (i.e., maximum possible value). (C) Light-propagation direction AB is neither parallel nor perpendicular to the optic-axis direction. An indicatrix

ordinary rays (n_ω). Either indicatrix may be represented by the mathematical formula

$$\frac{x^2 + y^2}{n_\omega^2} + \frac{z^2}{n_\epsilon^2} = 1.$$

For positive crystals, $n_\omega < n_\epsilon$ (i.e., the ordinary ray is the faster ray), and the spheroid is prolate; for negative crystals, $n_\omega > n_\epsilon$, and the spheroid is oblate; and when $n_\omega = n_\epsilon$, the spheroid becomes a sphere, and the crystal is isotropic.

Use of the Uniaxial Indicatrix

An indicatrix "indicates" two things: (1) the indices of refraction of the crystal for ordinary and extraordinary rays moving along a given path, and (2) the planes of polarization of waves moving along these two rays. An indicatrix actually tells no more than ray velocity surfaces do, but is much easier visualized.

Appropriately constructed and properly oriented with its optic axis (axis of rotation) parallel to c-axis, the indicatrix is imagined to lie within the crystal section or fragment that it represents in such a way that your line of sight passes through its center (Fig. 5-5). To view the crystal or fragment from any given direction, cut a section through the indicatrix normal to your line of sight—or more accurately, parallel to the plane tangent to the surface of the indicatrix at its intersection with your line of sight (Fig. 5-5,C). The section so derived is always an ellipse or a circle, which is merely a special form of an ellipse. The *lengths* of the semi-major and semi-minor axes of this ellipse represent the refractive indices of the two advancing rays, and the *directions* of the major and minor axes fit the planes of polarization of waves associated with the respective rays.

---

section perpendicular to *AB* yields only an approximation of vibration direction and refractive index for the extraordinary wave. A section parallel to *PQ* yields rigorous results for both ordinary and extraordinary waves, where *PQ* is tangent to the ellipsoid at *A* in the *AB*-optic-axis plane. The length of the semi-minor axis is n_ω, and the semi-major axis is n_ϵ' (i.e., some value between n_ϵ and n_ω). *Conclusion:* Light waves moving along *AB* are confined to vibrate parallel to the major and minor axes of the elliptical section. The ordinary wave vibrates parallel to the minor axis (i.e., always perpendicular to the optic axis) and experiences a refractive index of n_ω (the ordinary wave always has an index of n_ω). The extraordinary wave vibrates parallel to the major axis (*RS*) and experiences a refraction index of n_ϵ', intermediate between maximum (n_ϵ) and minimum (n_ω) values. *Note:* For a negative crystal (e.g., calcite) n_ϵ is less than n_ω, and the extraordinary wave will always vibrate parallel to the minor axis of the elliptical indicatrix section and experience a refraction index less than n_ω.

Looking down the optic axis (Fig. 5-5,A), you see a circular section whose radius (semi-axis) is n_ω, and the infinite possible circle diameters denote the infinite possible polarization planes of waves advancing parallel to the optic axis.

Viewing along a line perpendicular to the c-axis (Fig. 5-5,B), you see, normal to your line of sight, an elliptical section of maximum ellipticity called a principal section. The axis of this ellipse, which parallels the optic axis (i.e., major axis if the crystal is positive, and the minor axis if the crystal is negative), marks the vibration direction of the extraordinary wave, and one-half its length (i.e., the semi-axis) is n_ϵ, or the extreme refractive index for the extraordinary ray. The ellipse axis normal to the optic axis denotes the vibration direction of the ordinary wave, and the length of the semi-axis is n_ω, the unvarying refractive index for the ordinary ray.

Looking at an angle to the c-axis (Fig. 5-5,C), imagine an elliptical section through the center of the indicatrix parallel to a tangent to the indicatrix (PQ) at its intersection with your line of sight. Note that regardless where the section is cut, the ellipse always shows one axis that is normal to c and of length n_ω, again revealing the vibration direction and refractive index of the ordinary wave. The other axis of the elliptical section describes the extraordinary wave, its direction indicating the vibration direction of the extraordinary wave* and one-half its length representing the refractive index for the extraordinary ray. The refractive index for the extraordinary ray ranges from that of the ordinary ray n_ω to a maximum or minimum value n_ϵ. Only n_ϵ is of practical value; it is the index of only extraordinary rays moving perpendicular to the optic axis and is the only refractive index commonly measured for the extraordinary ray.

Although values in random section can be measured and calculated, the effort normally yields a rather hollow reward. For this reason, the procedure will not be discussed in this text.

The Nicol Prism

Even though most petrographic microscopes are equipped with polaroid disks, an examination of the optics of the Nicol prism offers an excellent exercise in the concepts just presented.

---

*Note that in random sections, extraordinary waves do not vibrate normal to their associated rays but vibrate in the elliptical section, which is tangent to the indicatrix surface at its intersection with the ray.

FIGURE 5-6. The Nicol prism. A calcite rhomb split diagonally and cemented with Canada balsam becomes an excellent polarizer.

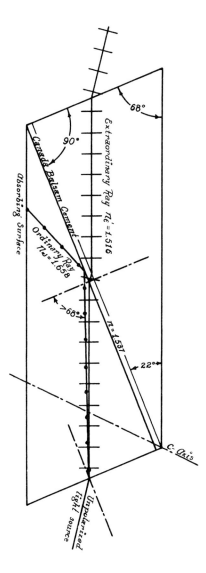

The Nicol prism is made from a rhomb of optically clear calcite measuring about 1 cm wide and 2 or 3 cm long. The rhomb is split as shown in Figure 5-6 and recemented with a layer of Canada balsam; the two ends are ground and polished at an angle slightly different from that of the natural rhomb; and the outer surface, excluding the two ends, is painted with a dull black, light-absorbing pigment.

To trace the light rays through the rhomb, consider a section that contains the c-crystallographic axis and is normal to the recemented plane (Fig. 5-6). A ray of unpolarized light incident on the lower surface of the calcite prism is refracted and split into two rays of plane polarized light. The ordinary ray, for which the calcite has the higher refractive index, is refracted more than the extraordinary ray, and strikes the balsam interface at an angle of incidence greater than the critical angle (CA).

$$\sin \text{CA} = \frac{n_{\text{balsam}}}{n_\omega} = \frac{1.537}{1.658} = 0.927, \qquad \text{CA} = 68°.$$

The ordinary ray, therefore, does not enter the low-index medium (i.e., the balsam) but is totally reflected at the balsam interface and is absorbed by the black paint on the side of the prism.

For the extraordinary ray, the calcite should exhibit a refractive index of about 1.516, which is less than the index of balsam. At the balsam interface, the extraordinary ray is refracted into the balsam toward the interface normal and then back to the original direction upon entering the upper calcite prism. Light waves moving along the extraordinary ray vibrate in the plane of the paper (i.e., in a plane parallel to the c-axis), and these emerge from the upper end of the prism to form an excellent source of plane polarized light.

CHAPTER 6

Uniaxial Crystals and the Petrographic Microscope

Interference Colors and Birefringence

In practice we examine mineral grains in rock thin sections or mineral fragments mounted on glass slides placed on the stage of a petrographic microscope. With uncrossed nicols the N–S polarized light from the lower nicol renders these fragments (or grains) visible by their relief or by their color. Between crossed nicols, anisotropic fragments show brilliant interference colors against a black background.

Plane polarized light emanating from the lower nicol may be represented as a N–S vector (Fig. 6-1) whose length represents the amplitude (i.e., intensity) of the polarized wave. Anisotropic mineral grains can only transmit light parallel to their inherent vibration directions, as revealed by an indicatrix section, and the single light vector is divided by the mineral into its two mutually perpendicular vectorial components parallel to these inherent directions. Each of these vectors has an E–W component that passes the upper nicol, causing the fragments to appear bright. At first glance, it may seem that these E–W components should cancel, since they are equal and opposite, but the two waves move through the mineral with different velocities and get out of phase. Interference caused by this phase difference produces interference colors.

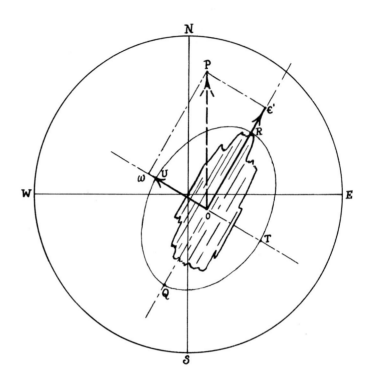

FIGURE 6-1. Vector representation of inherent vibration directions. Ellipse *RTQU* represents the appropriate indicatrix section for vertical light rays passing through the encircled crystal fragment, and the N–S vector *OP* represents the vibration direction and intensity (i.e., amplitude) of vertical light waves passing the lower nicol. Since the crystal fragment can pass only wave components that vibrate in planes parallel to its inherent vibration directions *RQ* and *TU*, vector *OP* may be divided into ordinary and extraordinary components O_ω and O_ϵ. As these two waves pass through the mineral fragment, they experience refractive indices of $n_\omega = OU$ and $n_\epsilon' = OR$, respectively. Note that the lengths of vectors O_ω and O_ϵ represent intensity, which changes with stage rotation. The lengths of vectors OU and OR represent refractive indices and do not change with stage rotation. Note also that this crystal is optically positive, since $n_\epsilon' > n_\omega$.

Extinction

Between crossed nicols, anisotropic mineral grains become dark, or *extinguished*, in certain orientations:

1. During a complete rotation of the microscope stage, a given mineral grain will become dark four times, at 90° intervals (Fig. 6-2). Whenever vibration directions of the mineral become N–S and E–W, the mineral allows the N–S wave from the lower nicol to pass through as a single wave, and no

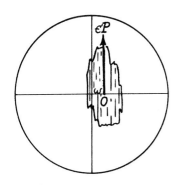 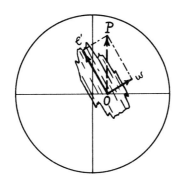 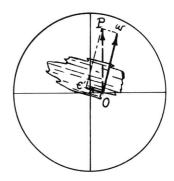

FIGURE 6-2. Extinction between crossed nicols. Every 90° of stage rotation brings one vibration direction (O_ω or O_ϵ) of the crystal into coincidence with the N–S vibration direction of the lower nicol. When the crystal is so oriented, the vector OP has no E–W component to pass through the E–W vibration direction of the upper nicol, and the fragment is dark between crossed nicols.

E–W component reaches the upper nicol. Brightness is at a maximum when mineral vibration directions are in 45° positions and at a minimum (zero) when N–S or E–W.*

2. Anistropic mineral particles should appear dark† when viewed along the optic axis (i.e., with the optic axis vertical), since the polarization direction of the incident light is unchanged.

Retardation

Two rays whose waves vibrate in mutually perpendicular directions pass through an anisotropic crystal with unequal velocities. The distance by which the slow ray lags behind the fast ray is called retardation (Δ), and is usually measured in millimicrons (mμ). If it takes the fast ray t_1 seconds and the slow ray t_2 seconds to move completely through a crystal grain, the retardation (Δ) is

$$\Delta = (t_2 - t_1)v,$$

---

*As light intensity approaches zero, interference colors do not change.

†Minerals with appreciable birefringence never appear completely dark, even when the optic axis is vertical, because light rays from the polarizer pass through a converging lens, and only the center ray is really parallel to the optic axis.

where v is the velocity of light in air. Since the thickness of the crystal (d) equals $V_1 t_1$ or $V_2 t_2$,

$$t_1 = \frac{d}{V_1}$$

and

$$t_2 = \frac{d}{V_2}$$

or, since $V_1 = v/n_1$ and $V_2 = v/n_2$,

$$\Delta = (t_2 - t_1)v = \left(\frac{d}{V_2} - \frac{d}{V_1}\right)v = d\left(\frac{n_2}{v} - \frac{n_1}{v}\right)v$$

or

$$\Delta = d(n_2 - n_1).$$

Birefringence

The quantity ($n_2 - n_1$), or the *refractive index of the anisotropic medium for the slow ray minus the refractive index for the fast ray, is called the birefringence.* Birefringence is dependent upon:

1. The mineral species. Some minerals, like leucite, have a maximum birefringence of 0.001 or less, whereas others, like calcite or rutile, have a maximum birefringence of 0.3 or more.

2. The direction of observation. *Birefringence of a uniaxial crystal varies from zero for light propagated parallel to the optic axis to some maximum value for light propagated normal to the optic axis (See Fig. 5-5). Maximum birefringence is a fixed characteristic of the mineral variety.*

3. The wavelength of the incident light. Wavelength affects the birefringence only slightly; when a numerical value is given, it is usually assumed to be for Na_D radiation.

Phase Difference and Interference

Consider a fragment of a uniaxial positive crystal (i.e., one in which the ordinary ray is the faster ray) represented by the prominent box in Figure 6-3; for simplicity, the box is oriented with the *c*-axis horizontal. A wave of *monochromatic* light vibrating in a N–S plane emanates from the lower nicol; upon entering the crystal, this wave is divided vectorially into its two mutually perpendicular components, which vibrate parallel to the inherent vibration directions of the mineral and form ordinary and extraordinary waves. These two waves move through the crystal with unequal velocities, and hence unequal wavelengths ($V = f\lambda$). In Figure 6-3,A, the ordinary and the extraordinary wave both complete some whole number of wavelengths, or vibrations (the extraordinary wave, having the shorter wavelength, completes more vibrations than the ordinary wave), in passing completely through the crystal of thickness d; and the retardation is $n\lambda$, where n is a whole number. Although both ω and ϵ waves have E–W components, they are equal and opposite and nullify one another, so that no light passes the upper nicol, and the crystal is dark. We may then conclude that *when the retardation is $n\lambda$ (i.e., when the waves are 0°, 360°, etc., out of phase), we observe minimum light intensity.**

Figure 6-3,B shows a situation in which the ordinary wave completes $n\lambda$ vibrations and the extraordinary wave completes $(n + \frac{1}{2})\lambda$ vibrations in passing through the crystal. Again both waves have E–W components, but they now have the same direction and add to produce a maximum E–W component that passes the upper nicol, making the crystal of thickness d appear bright under crossed nicols. Therefore, *when the retardation equals $(n + \frac{1}{2})\lambda$ (i.e., when the waves are 180°, 270°, etc., out of phase), we observe maximum light intensity.*

The amount of light transmitted by the upper nicol is obviously a function of the phase difference between the two waves transmitted by the crystal, and is proportional to $\sin^2(\Delta/\lambda)180°$, where Δ/λ is the phase difference expressed as a fraction of the incident wavelength. When the retardation is $n\lambda$, the amount of light passing the upper nicol is zero, and when $\Delta = (n + \frac{1}{2})\lambda$ the amount of light passing the upper nicol is maximum (Fig. 6-4,A).

Between parallel nicols, conditions are reversed, and maximum light from the crystal passes the upper nicol when $\Delta = n\lambda$, and zero light when $\Delta = (n + \frac{1}{2})\lambda$; the amount of light that passes the N–S analyzer is proportional to $\cos^2(\Delta/\lambda)180°$ (Fig. 6-4,B).

The Interference Color Sequence

If monochromatic light is replaced by white light, then as retardation is changed, a certain sequence of interference colors appears in place of the alternating light and dark bands (Plate III). Assuming uniform dispersion (i.e., $n_2 - n_1$, for red, $= n_2 - n_1$, for violet), retardation $[\Delta = d(n_2 - n_1)]$ is

---

*The student should recall that these conditions obtain only when monochromatic light is used.

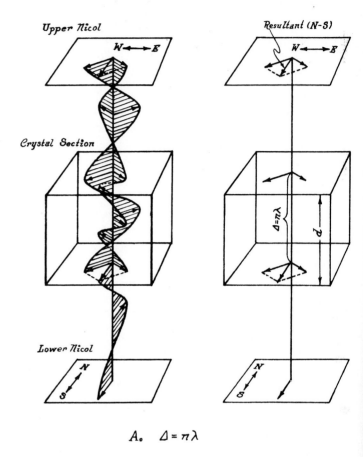

A. $\Delta = n\lambda$

FIGURE 6-3. Phase relations. A *monochromatic* wave of polarized N–S light from the lower nicol is split into ordinary and extraordinary waves by the crystal. These waves vibrate in mutually perpendicular planes, advance with different velocities, and have different wavelengths. Upon emerging from the crystal, the two waves resume their original velocity and wavelength, but are now out of phase by an amount equal to the retardation $\triangle$. If the retardation allows the two waves to proceed without apparent phase change, the E–W components of the two waves are equal and opposite (i.e., their resultant is N–S) and no light passes the upper nicol. If the retardation causes the two waves to proceed $\frac{1}{2}\lambda$ out of phase, the E–W components of the two waves add to give an E–W resultant and maximum light intensity passes

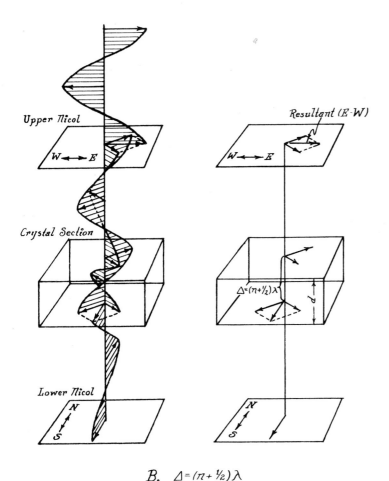

B. $\Delta = (n + \frac{1}{2})\lambda$

the upper nicol. (A) Minimum light passes the upper nicol when retardation within the crystal is some full multiple of the wavelength ($\triangle = n\lambda$). The left diagram shows that the wave of long wavelength completes one full wavelength and the wave of short wavelength completes two full wavelengths within the crystal. The right diagram shows only essential vectors.
(B) Maximum light passes the upper nicol when retardation is some full multiple of the wavelength plus one-half ($\triangle = n + \frac{1}{2}\lambda$). The left diagram shows that the wave of long wavelength completes one-half its wavelength and the wave of short wavelength completes one full wavelength within the crystal. The right diagram shows only essential vectors.

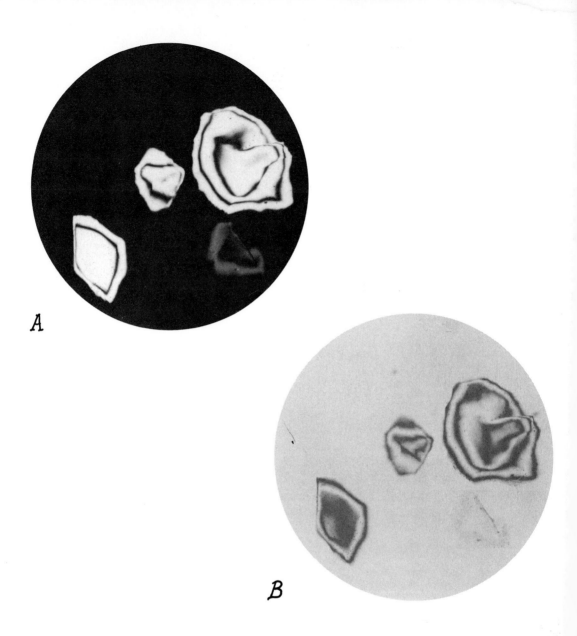

FIGURE 6-4. Quartz grains in monochromatic (sodium D) light between crossed nicols and parallel nicols. (A) Between crossed nicols, a crystal fragment appears as successive bands of brightness (yellow) and darkness with increasing thickness. A fragment of given orientation has constant birefringence, and retardation of waves vibrating parallel to its inherent vibration directions increases uniformly with thickness [$(\triangle = t(n_2 - n_1)$]. Maximum brightness appears where $\triangle = \frac{1}{2}\lambda, \frac{3}{2}\lambda, \frac{5}{2}\lambda$, etc., and maximum darkness where $\triangle = \lambda, 2\lambda, 3\lambda$, etc. (B) Between parallel nicols, the crystal fragment again appears as successive light and dark bands, but maximum brightness now appears where $\triangle = \lambda, 2\lambda, 3\lambda$, etc., and maximum darkness where $\triangle = \frac{1}{2}\lambda, \frac{3}{2}\lambda, \frac{5}{2}\lambda$, etc.

the same for every wavelength at a given crystal thickness. A given retardation represents $n\lambda$ for some wavelengths, $(n + \frac{1}{2})\lambda$ for others, and intermediate values for still other wavelengths; the combination of wavelengths passing the analyzer nicol produces a given interference color for each given retardation.

This particular sequence of interference colors was noted long ago by Sir Isaac Newton in his study of thin films, and can be observed by any student on a rainy day when thin films of oil appear on the wet asphalt of the campus parking lot. To derive this sequence, consider a thin uniform wedge of quartz cut with its c-crystallographic axis in the plane of the wedge normal to its length. For light rays moving normal to the plane of the wedge, the birefringence is fixed $(n_\epsilon - n_\omega)$, and the retardation at any point along the wedge depends only on its thickness. Using monochromatic sodium light ($\lambda = 589$ mμ), we could examine such a wedge oriented in the 45° position on our microscope stage. With crossed nicols we would observe that alternating bands of yellow light and darkness grade into each other as the thickness changes (Plate III,A).* At the thin end, the thickness and hence the retardation are zero, and the wedge is dark. With increasing thickness, the retardation increases, and the wedge brightens to a maximum when the retardation equals 295 mμ (i.e., $\frac{1}{2}\lambda$). Further increase causes light intensity to decrease to zero when $\Delta = 589$ mμ (i.e., λ), and to increase again to maximum brightness when $\Delta = 1\frac{1}{2}\lambda$, etc., as shown in Plate III,A. If monochromatic red light (e.g., $\lambda = 700$ mμ) were used (Plate II), the alternating dark and light bands would be more widely spaced, since the first bright (red) band would be centered at $\Delta = 350$ mμ ($\frac{1}{2}\lambda$); if monochromatic violet light were used (e.g., $\lambda = 400$ mμ), the alternating dark and bright (violet) bands would be more closely spaced, since the first bright band would be centered at $\Delta = 200$ mμ ($\frac{1}{2}\lambda$). Since white light contains many wavelengths between red and violet, we may add color bands of representative wavelengths (Plate II) to see what effect will be observed under crossed nicols when the quartz wedge is illuminated by white light. Where the thickness is near zero, there is no color; with increase in thickness, black grades into violet-gray, and then all colors begin to appear, long wavelengths appearing last. At a thickness of about 0.02 mm (i.e., 200 mμ retardation) all colors are transmitted with moderate intensity and add to form white light, which grades into yellow, orange, red, violet, blue, and green as their complementary colors—violet, blue, green, yellow, orange, and red—successively extinguish. The color sequence, begin-

---

*The student is encouraged to confirm this simple observation by using the quartz wedge accessory that should accompany his microscope.

ning with yellow and ending with green, repeats over and over again with increase in wedge thickness, but with each repetition the colors become less brilliant, pale pinks and greens predominating until all colors blend into a uniform, "dirty," high-order white. Near the thin edge of the wedge, all wavelengths are seen in a continuous range, and combine to form one brilliant color, but with increasing thickness the colors get "out of phase" and confused, and wavelengths from many parts of the spectrum combine to form high-order white.

ORDERS OF INTERFERENCE COLORS. For convenience in discussion, the sequence of interference colors is divided into equal units called orders, each encompassing 550 mμ of retardation and separated by a red color bar. At the end of the first order of colors ($\Delta = 550$ mμ) is a very narrow band of reddish-violet called first-order red, or *sensitive violet*, because of the noticeable color change that is produced by even a small change in retardation.

High-order colors are difficult to distinguish, since they are all pastel shades of pinks and greens grading ultimately to shades of high-order white. It is often possible, however, on a wedge-shaped edge of a mineral grain to count the number of orders to the upper grain surface.

For the beginning student, there is often confusion in distinguishing minerals of very low birefringence, which show first-order white, from minerals of very high birefringence, which show high-order white. Experience will soon eliminate confusion, but in the interim a few suggestions may be welcome. First-order white appears as an even white, ranging from bluish-gray in small fragments and for near-axis orientations to yellowish-white in large fragments and for orientations normal to the optic axis. Large fragments may even show vivid first-order yellow, orange, and red on their thickest parts. High-order white is less uniform, often showing flecks of color; particle size has little effect on color. A decisive test can be made with the gypsum plate. When the plate is inserted, the additional retardation it produces will have little noticeable effect on crystals with high birefrigence, but crystals with low birefringence assume vivid colors of first-order yellow or second-order blue. Rotation of the upper nicol from its crossed position will also cause only low-birefringent crystals to change color.

INTERFERENCE COLOR SEQUENCE (PARALLEL NICOLS). Between parallel nicols, anisotropic crystal plates or fragments show interference colors against the white background of an illuminated field (Plate III,D). Since the analyzer nicol now passes those wavelengths that are out of phase by $n\lambda$ and absorbs those with $(n + \frac{1}{2})\lambda$ phase relations, the colors in the sequence are essentially the complements of those in the normal sequence (Plate III,C).

Function of Accessory Plates

Construction of the *quartz wedge* (Fig. 2-15,C) was described on page 40 and its optical function in the previous section. Since quartz is optically positive ($n_\epsilon = 1.5533$ and $n_\omega = 1.5442$), the extraordinary ray is slow and vibrates parallel to the optic axis, as marked on the frame. The wedge has a uniform birefringence of 0.0091 ($n_\epsilon - n_\omega$) and ranges in thickness from essentially zero to about 0.25 mm. Its sole function is to produce retardation ranging from 0 mμ to a high order.

The *gypsum (first-order red) plate* (Fig. 2-15,A) is a uniform plate of gypsum or quartz about 0.0625 mm thick used to produce 550 mμ retardation equivalent to the sensitive violet. The arrow marked on the metal frame indicates the polarization direction of the slow wave, which lies in the plane of disk, perpendicular to the length of the frame.* The gypsum plate is inserted into the microscope accessory slot with its direction of slow vibration (γ) parallel to one inherent vibration direction of the crystal grain being viewed. If the slow wave of the plate parallels the slow wave of the crystal, it is retarded by both crystal and gypsum plate, and the total retardation equals the sum of the two separate retardations (i.e., the retardation caused by the crystal plus 550 mμ). If the slow wave of the gypsum plate parallels the fast wave of the crystal, however, the net retardation is the difference between the two separate retardations (i.e., the retardation caused by the crystal minus 550 mμ); thus what was originally the fast ray may end up the slower ray after passing the gypsum plate.

Students may wish to learn this rule of thumb: *Slow wave parallel to slow wave causes addition of retardation, and the "colors add"; slow wave parallel to fast wave causes subtraction of retardation, and the "colors subtract"* (Plate VIII).

The *mica plate*, or $\frac{1}{4}\lambda$ *plate* (Fig. 2-15,B) is the third common accessory plate. Its function is to cause the wave vibrating parallel to the arrow on the metal frame to lag its faster companion by about 147 mμ (i.e., one-fourth the sodium wavelength), which is equivalent to first-order white. It is used in the same manner as the gypsum plate, and causes interference colors to "add" or "subtract" by 147 mμ.

---

*The circular disks of gypsum or mica are often poorly cemented in their frames and may become rotated, which renders the accessory useless and leads to endless frustration for the beginning student. The student is therefore encouraged to rotate the accessory on the microscope stage between crossed nicols to see that the vibration directions are, in fact, parallel and normal to the indicated direction and to check the relative velocities of the two rays with another accessory plate.

The Birefringence Chart

The most useful single illustration in this book is the birefringence chart, opposite this page. From left to right the chart displays the normal sequence of interference colors for crossed nicols as previously derived for the quartz wedge (Plate III,C). Each color results from a specific retardation, indicated in millimicrons (mμ) across the bottom of the chart. Retardation, and hence color, is dependent upon thickness, represented by horizontal lines labeled in millimeters on the left, and birefringence ($n_2 - n_1$), represented by lines of equal birefringence radiating from the lower left and labeled across the top and right side of the chart. The chart is simply a graphical representation of the formula $\Delta = d(n_2 - n_1)$ and allows determination of the numerical value of any one of the three variables if the other two are known.

For example, rock thin sections are ground to the preferred thickness (0.03 mm on left ordinate) by making use of the known maximum birefringence of a prominent mineral, usually quartz (0.009 at top of chart), and the appropriate first-order interference color ($\Delta = 275$ mμ on vertical line). When the straw yellow color that quartz shows at a thickness of 0.03 mm appears, grinding is stopped.

The Quantitative Measure of Birefringence

Only the thickness of a given crystal grain and its retardation (interference color) under crossed nicols are needed to determine its numerical birefringence. The student must, however, keep several basic principles constantly in mind in order not to be misled by the apparent simplicity of this procedure. He must be constantly aware that the numerical birefringence of any given crystal can vary from zero to some maximum value, depending upon the direction of observation; *only the maximum value is a characteristic property of the mineral species.* In rock thin sections, where mineral grains are of uniform thickness (0.03 mm), the maximum birefringence is determined on the basis of a grain that shows maximum interference colors (i.e., maximum retardation). The grain must, of course, be selected from a large number of randomly oriented grains. Unfortunately, thin sections are often thicker than the standard of 0.03 mm, and are usually thinnest near the edges. A student can, however, learn to estimate maximum birefringence with sufficient accuracy to make the effort worthwhile.

Crystal fragments of varying size and orientation present even more complications, and birefringence estimates are correspondingly less accurate, but, again, the results can richly reward the effort. The student should be aware of

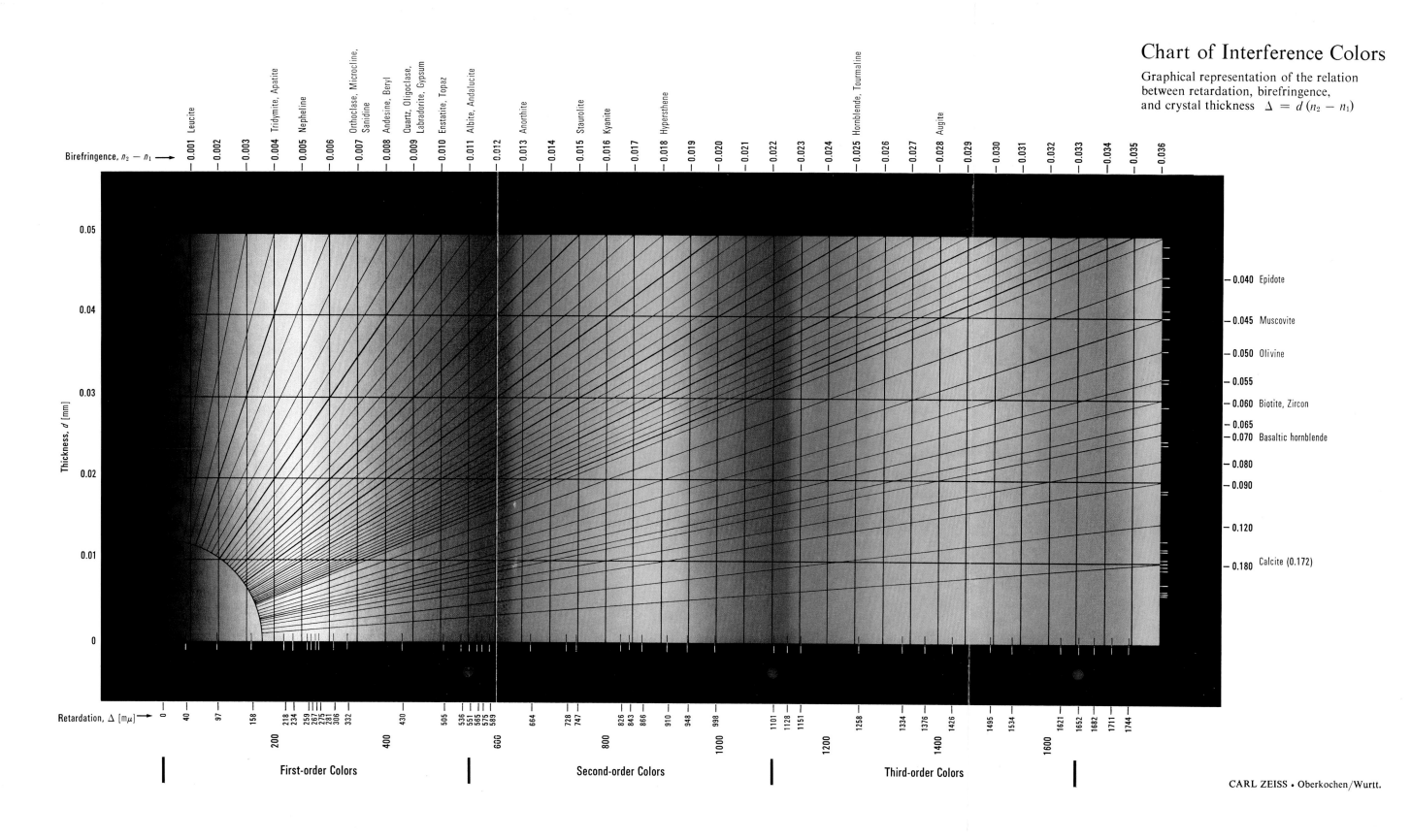

two obvious considerations. (1) Loose fragments consistently lie on their broad side, and it is safe to assume that fragment thickness is slightly less than the narrow dimension that he can see and measure (see p. 23). (2) A small fragment has a higher birefringence than a large fragment that shows the same retardation. A logical procedure, then, is to look for a relatively small fragment that shows high retardation and estimate its thickness. Try it several times.

Several simple microscope accessories have been designed to facilitate the measurement of numerical birefringence, notably the Berek and slot compensators. Each consists basically of a crystal plate that can be inserted in the microscope accessory slot and rotated so as to increase its retardation in opposition to the retardation of the mineral being examined. The amount of rotation necessary to reduce the retardation of the mineral to zero affords an accurate measure of retardation, which can be read directly on the scale of the compensator. Unfortunately, the measurement of maximum birefringence requires an accurate knowledge of thickness and crystal orientation in addition to retardation; of the three, retardation can be visually estimated with greatest accuracy.

Birefringence of Uniaxial Minerals

We have seen (see p. 82) that calcite has large negative birefringence* owing to strong atomic polarization in the plane of the CO_3^{--} ion, and we may extend this principle to include all uniaxial crystals containing a planar molecular unit of highly refracting anions (e.g., usually O^{--}). All trigonal carbonates show high "negative birefringence." Uniaxial nitrates also show high negative birefringence due to their planar nitrate ions (NO_3^-).

To extend this principle to other uniaxial minerals we need only recall that high atomic polarization leads to high index of refraction, or waves vibrating parallel to the directions of high polarization travel with low velocity. The shape of complex polarized ions and their orientation in the crystal structure is, therefore, the key to predicting birefringence.

Some highly polarized diatomic ions, such as the carbide ion (C_2^{--}) and the peroxide ion (O_2^{--}), are elongated like frankfurters and are oriented parallel to the unique crystallographic axis (*c*-axis) to yield tetragonal, trigonal, or hexagonal crystals. Light waves moving along *c* vibrate in opposition to the ion polarization and move with relatively high velocity (i.e., n_ω is small), and

---

*Some authors use the expression "negative birefringence" if the mineral is optically negative (i.e., if $n_\omega > n_\epsilon$) and "positive birefringence" for the reverse condition.

light waves moving normal to c and vibrating parallel to c (i.e., in the direction of the extraordinary wave) increase the polarization and move with a relatively low velocity (i.e., n_ϵ is large). Although carbides and peroxides do not form important mineral groups, the relation between their structure and birefringence suggests that other uniaxial crystals containing such diatomic ions may also have high positive birefringence.

Highly polarized units like the silicate ion (SiO_4^{----}), phosphate ion (PO_4^{---}), sulfate ion (SO_4^{--}), and chromate ion (CrO_4^{--}) consist of a small cation surrounded by four highly refractive anions that occupy the corners of a regular tetrahedron; in spite of the great polarization between cation and anions, however, these units are as a whole basically isotropic due to their high-order symmetry. The birefringence of crystals containing these units is dependent upon the way in which they and the other constituents of a particular crystal structure are arranged, and may vary from zero to a high value. Uniaxial phosphates, sulfates, and chromates tend to show low birefringence. Maximum birefringences of silicate minerals are extremely variable because of the great variety of structural types. Chain and sheet silicates have definite, unique structural directions and tend to show high birefringence (e.g., pyroxenes $\simeq 0.025$, amphiboles $\simeq 0.025$, micas $\simeq 0.04$, talcs $\simeq 0.04$); framework silicates tend to approximate isotropic structures and usually exhibit low birefringence (e.g., quartz 0.009, feldspars $\simeq 0.008$, feldspathoids $\simeq 0.003$, zeolites $\simeq 0.006$); and orthosilicates may show either strong or weak birefringence, depending on the arrangement of independent SiO_4^{----} units and intervening cations (e.g., garnet 0.000, olivine $\simeq 0.04$, topaz $\simeq 0.009$, zircon $\simeq 0.06$).

Anomalous Interference Colors

Several anomalous or abnormal interference phenomena are fairly common; for example, an isometric crystal structure may be sufficiently distorted by mechanical stress to show obvious double refraction.* Although the resulting interference colors normally exhibit the usual color sequence, the phenomenon is considered anomalous.

Many minerals are deeply colored and act as color filters by absorbing certain wavelengths. Deep green minerals, for example, largely absorb red, which appears as an anomalous, dark band in the normal interference sequence.

Usually, however, the term "anomalous" is applied to color sequences that depart from the sequence shown in the color chart, and which are related in

---

*Several phases of industry take advantage of this fact to reveal the presence of strain and its effects on normally isometric crystalline substances

several possible ways to dispersion. It is of course inherent in the concept of dispersion that different indices of refraction obtain for different wavelengths, and a logical extension is that birefringence will differ with wavelength. Birefringence for one end of the visible spectrum may be twice that for the opposite end, causing a shift in the position of dark and bright bands for certain colors and resulting in abnormal color effects. Some minerals are essentially isotropic (i.e., have low birefringence) for some wavelengths yet have appreciable birefringence for others, which means that certain wavelengths are absent from the normal color sequence. Several rather common uniaxial minerals (e.g., idocrase and melilite) are essentially isotropic for wavelengths near the center of the visible spectrum in that they are deficient in orange, yellow, and green. First-order gray of the normal color sequence becomes a deep indigo blue in the absence of the middle wavelengths; either a dirty yellow or a pale olive drab appears in place of first-order white and yellow, and grades directly into red-violet with increasing thickness. Such minerals have low birefringence for the entire spectrum; deep indigo blue and pale olive drab are the interference colors commonly observed.

Uniaxial Interference Figures

Conoscopic Illumination

The substage condensing lens system is so designed that its numerical aperture can be changed from about 0.2 to nearly 1.0 either by raising the position of a single condensing lens or by inserting a second, movable, substage condensing lens. Although the light rays are always convergent, the convergence is considerably less when only the lens of low N.A. is used, and the light rays are said to produce *orthoscopic illumination* (Fig. 6-5,A), a term that suggests the nearly parallel disposition of the illuminating rays. An appreciable increase in the N.A. of the condensing lens system can be attained by rotating the auxiliary condensing lens into the light path. This causes the illuminating rays to be highly convergent so that they come to a point focus and cross within a viewed crystal grain; such a lens arrangement is said to produce *conoscopic illumination* (Fig. 6-5,B), suggesting the convergent or cone-like nature of the illuminating rays. Upon emerging from the crystal the light rays diverge rapidly, and an objective lens of similarly high N.A. (i.e., a high-power objective lens) is necessary to interpret the emergent cone of light.

Under conoscopic illumination with crossed nicols, the interfering light rays that emerge from a crystal grain form useful light patterns, called interference

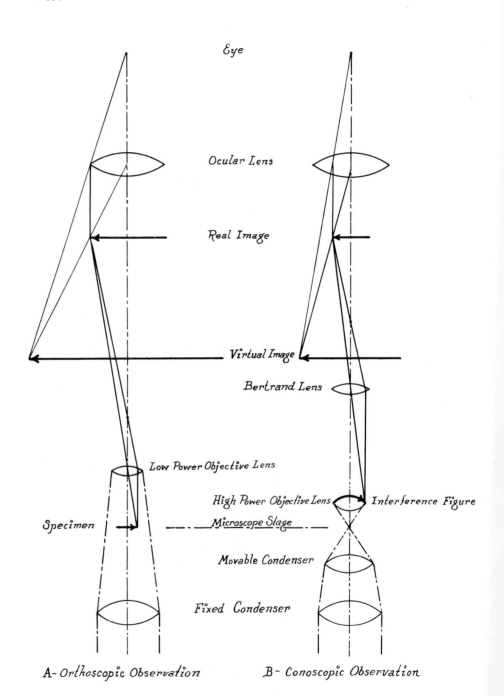

FIGURE 6-5. Orthoscopic and conoscopic illumination. (A) Orthoscopic illumination is normal illumination; the ocular and objective lenses form a simple microscope, and the object in focus lies on the microscope stage at the focal point of the objective lens. The fixed condenser supplies a cone of mildly convergent light just large enough to fill the objective. (B) Conoscopic illumination requires strongly convergent light from the movable condenser to fill the high-power objective lens. The ocular and Bertrand lenses form a simple microscope, and the object in focus is the upper surface of the high-power objective lens, where the interference figure is formed. The same result can be achieved without magnification by removing the ocular and Bertrand lenses and viewing the upper surface of the objective directly.

figures, on the upper surface of the objective lens. Interference figures may be viewed in two ways. By removing the ocular and looking through the microscope tube directly at the objective lens, a small but distinct figure can be seen. By inserting the Bertrand lens, which brings the upper surface of the objective into the focal plane of the ocular, a magnified figure can be seen. (If the cone of light does not completely fill the objective lens, only a small spot of light will appear. Raising the entire substage will enlarge the figure to fill the objective.)

To summarize, interference figures are viewed with the microscope arranged for conoscopic observation as follows:

1. Increase the N.A. of the substage condensing lens system by inserting the swing-in auxiliary condenser or raising the single condensing lens.

2. Focus the high-power objective lens on the crystal grain to be viewed.

3. Cross the nicols.

4. Remove the ocular or insert the Bertrand lens.

Isochromatic Color Bands (Isochromes)

An interference figure is a pattern of bright areas and black bars or bands. The bright areas often contain rings of interference colors called *isochromes*; the black bars or bands, called *isogyres*, are formed where vibration directions in the figure coincide with polarization directions of the microscope nicols.

Consider a crystal plate of a uniaxial mineral placed on the microscope so that its *c*-crystallographic axis is oriented normal to the stage (Plate VI). Although the optic axis is vertical, only the central ray traverses the crystal in a direction parallel to the optic axis because all others are convergent. Birefringence of the crystal for waves moving along this ray is of course zero, and the center spot of the interference pattern is black, because the retardation must also be zero. This dark spot of zero retardation, which marks the "emergence of the optic axis," is called the *melatope*.

All rays other than the central one converge to a focal spot within the crystal plate and diverge upward. Inclination is slight for rays near the central path but increases rapidly with distance from the central ray. Rays inclined to the optic axis mark transmission directions of the ordinary waves, which vibrate normal to c, and the extraordinary waves, which vibrate in a principal indicatrix section that contains c. Retardation between ordinary and extraordinary waves produces interference colors. Since both the thickness and the birefringence of the mineral plate increase with the inclination of rays, the waves that pass through the crystal plate emerge in cones of equal retardation, varying from zero at the melatope to a maximum value at the edge of the field. The interference figure thus consists of circles of equal retardation (i.e., of equal color), called isochromatic bands or isochromes. From first-order black at the melatope, the isochromes progress through the normal interference color sequence (Plate VI). The number of these color bands, or the maximum retardation for peripheral waves, depends both on the thickness and the maximum birefringence of the mineral. For a given mineral, thick sections show more isochromatic color rings than do thin sections, and a strongly birefringent mineral will show more isochromes than a mineral variety having weak birefringence. Although retardation, and hence color, is directly proportional to both thickness and birefringence, the latter property has the greater influence upon the number of isochromes produced. Since the maximum birefringence of calcite, for example, is 22 times that of quartz, a quartz plate 22 times as thick would be necessary to produce equivalent isochromes. Seldom do we deal with sections or even fragments that vary in thickness by more than $3\times$ or $4\times$ and, hence, great numbers of isochromes usually indicate large birefringence.

Isogyres

Superimposed upon the isochromatic rings is a black cross made up of N–S and E–W zones of extinction called isogyres. These intersect at the melatope. Isogyres are not an interference phenomenon but result where planes of polarization of advancing waves parallel those of polarizer and analyzer. (Fig. 6-6). All extraordinary waves vibrate in a principle section of the indicatrix and hence are radial to the optic axis. Vibration directions of the extraordinary waves are inclined to the plane of the figure and are not perpendicular to c. Ordinary waves vibrate normal to extraordinary waves, normal to the optic axis direction and hence tangent to the isochromatic circles. Along the N–S

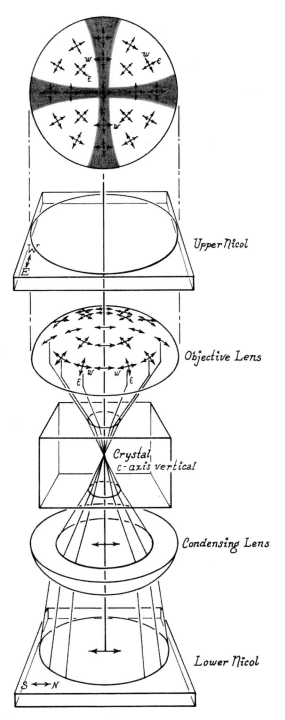

FIGURE 6-6. Formation of isogryres. As light waves from the lower nicol enter the crystal, each N–S wave is divided into mutually perpendicular ordinary (ω) and extraordinary (ϵ) waves that parallel the inherent vibration directions of the crystal. Since N–S waves are eliminated by the lower nicol and E–W waves by the upper nicol, an isogyre cross is formed where inherent crystal vibration directions parallel those of the microscope nicols. The c-axis of the uniaxial crystal is assumed to be vertical; the optic axis is thus the *direction* normal to the plane of the upper circle. Vibration directions of ordinary waves must be normal to the optic axis and form tangents to circles centered on the melatope. Vibration directions of extraordinary waves form a radial pattern from the melatope and *do not lie in the plane of the figure.*

and E–W axes of the interference figure, polarization planes lie N–S and E–W. Since the polarizer passes no E–W components and the analyzer passes no N–S components, dark bands of extinction form the uniaxial cross.

Isogyres show finite width and widen, or "fan out" near the edge of the field. A detailed explanation of this fanning out is given by Kamb (1958), but let it suffice here to note that polarization directions nearly parallel those of the nicols for a wide isogyre near the field periphery and for a very narrow isogyre near the melatope. The fanning out of isogyres is much more prominent in figures showing many isochromes than in figures showing few.

The student will soon discover that conoscopic illumination and crossed nicols produce a faint image resembling the uniaxial cross even when no crystal is being viewed. This results from a slight rotation of N–S waves by the almost spherical upper surface of the lowest element of the objective lens. These pseudo-isogyres are very weak and should not interfere with observation of the true ones.

Types of Interference Figures and Their Optic Signs

OPTIC AXIS FIGURE (*Fig. 6-7,A*). The interference figure described in the preceding sections is an optic-axis figure that forms only when the optic axis is vertical. Slight inclination of the optic axis (Fig. 6-7,B) causes the melatope to be slightly displaced from the center of the field. As the stage is rotated, the melatope moves in a circle about the center of the field in the same direction as the stage is rotated. Isogyres always intersect at the melatope and always remain N–S and E–W; during stage rotation they move back and forth parallel to the cross hairs of the ocular as the melatope subscribes its circle

FIGURE 6-7. Isogyre figures of uniaxial crystals. The appearance of a uniaxial interference figure is governed by the orientation of the single optic axis. Vibration directions are represented here by "longitude-latitude" lines on an appropriate indicatrix. Projection of these vibration directions to an interference figure delineates the isogyres. (A) a centered optic axis figure is formed when the optic axis is vertical; the melatope lies at the center of the field. Vibration directions of ordinary waves (ω) are concentric about the melatope, and extraordinary waves (ϵ) are radial from it. Isogyres form where vibration directions are N–S or E–W. (B) A slightly off-center optic-axis figure is formed when the optic axis is slightly inclined to the vertical. Vibration directions remain concentric and radial about the melatope. (C) The melatope moves beyond the visible field as the optic axis is further inclined, and only one isogyre may be visible at a time. (D) A flash figure is formed when the optic axis is horizontal. Vibration directions tend to be uniform throughout the visible field; the field is largely dark when the optic axis is N–S or E–W and entirely bright when the optic axis is rotated a few degrees from these directions in the horizontal plane.

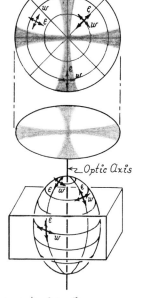

A - Optic Axis Figure (centered)

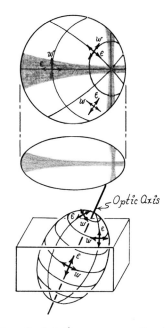

B - Optic Axis Figure (slightly off center)

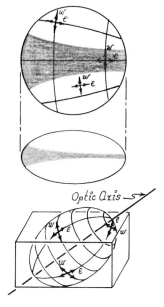

C - Optic Axis Figure (far off center)

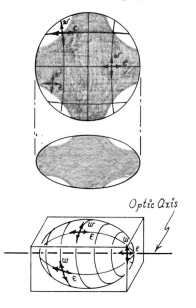

D - Flash Figure

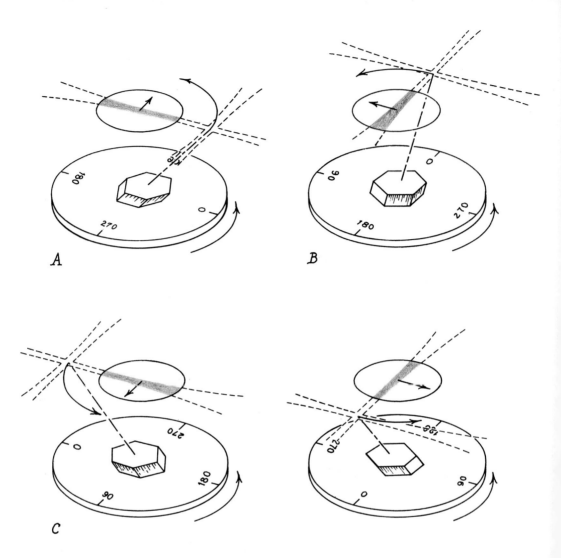

FIGURE 6-8. Isogyre movement with stage rotation. A centered optic-axis figure shows no movement with stage rotation, but the four arms of an off-center figure sweep successively across the visible field parallel to N–S or E–W directions. The melatope lies somewhere beyond the visible field, and by viewing the motion of two successive arms, one should be able to determine where it is and which quadrant of the uniaxial cross lies within the visible field. (A) The optic axis is inclined to the right, causing the melatope to be east of the visible field, and only the west arm of the isogyre cross is visible. As the stage is rotated counterclockwise, the melatope circles the visible field in the same direction. The E–W isogyre moves north and disappears from the field, and a N–S isogyre enters from the east and moves westward. (B) The melatope is north of the field, and the south arm of the cross is visible. Additional counterclockwise rotation causes the N–S isogyre to move westward beyond the field and the E–W isogyre to return moving southward. (C) The optic axis is now inclined to the left; the melatope is west of the field, and the east arm of the isogyre divides the visible field. (D) Another 90° rotation inclines the optic axis toward the observer; the melatope is south of the field, and the north arm of the cross is visible.

(Fig. 6-8). Further inclination of the optic axis causes the melatope to be further displaced, but so long as the melatope remains within the visible field, the interference figure may be called an optic-axis figure and will very adequately serve to reveal the uniaxial nature of the crystal and its optic sign.

Suppose that a uniaxial optic-axis figure formed by a crystal plate of low maximum birefringence has no isochromatic rings and that the entire field beyond the isogyres is first-order white (Plate IV,A or B), indicating a retardation of about 150 mμ.*

The gypsum plate, when inserted into the accessory slot, will cause the white field to become first-order yellow ($\Delta = 400$ mμ) in two opposite quadrants of the cross and second-order blue ($\Delta = 700$ mμ) in the remaining two (Plate IV,A). Recall that the sole function of a gypsum plate is to retard the wave vibrating parallel to the arrow on the mount (i.e., the slow ray) 550 mμ with respect to the wave vibrating normal to it. Since the arrow is oriented NE–SW when the plate is in use, the slow vibration direction of the gypsum plate parallels the ordinary wave in the NW and SE quadrants, and the extraordinary wave in the NE and SW quadrants of the cross. If the ordinary ray is the faster ray (i.e., the crystal is optically positive), it must lead the extraordinary ray by about 150 mμ upon emerging from the crystal plate to produce first-order white. In passing through the gypsum plate, the ordinary ray is slowed 550 mμ in the NW and SE quadrants, and the extraordinary ray is retarded 550 mμ in the NE and SW quadrants. The resulting retardation is 400 mμ (i.e., 550 mμ − 150 mμ), or first-order yellow, in the NW and SE quadrants and 700 mμ (i.e., 550 mμ + 150 mμ), or second-order blue, in the NE and SW quadrants (Plate IV,A). If the ordinary ray is slower (i.e., if the crystal is optically negative), the quadrant colors are reversed. The student should concentrate on one quadrant at a time, say the NE, or upper-right, forgetting the rest of the figure; addition of retardation (e.g., second-order blue) indicates an optically positive crystal and subtraction of retardation (e.g., first-order yellow) indicates an optically negative crystal. Since the resultant colors are a function of retardation within the crystal (i.e., initial interference colors), it is important for the student to be aware that he is not looking specifically for blue or yellow but is trying to determine where addition or subtraction of retardation has occurred. When an interference figure is produced by a crystal of great inherent birefringence, many isochromatic rings appear in the figure, and the addition or subtraction of 550 mμ of retardation may be negligible compared to the retardation already present. In this event, the quartz wedge is a more useful

---

*This value is necessarily greater near the periphery, but the white of the field may nevertheless appear quite uniform.

accessory than either the gypsum or the mica plate. Even when many isochromes are present, however, the gypsum plate often produces tiny spots of blue and yellow immediately adjacent to the melatope, in place of the tiny ring of first-order white.

The black isogyres, which represent zero retardation, become first-order red ($\Delta = 550$ mμ) with the gypsum plate.

The mica plate is used in the same manner as the gypsum plate (Plate IV,B) to retard by 147 mμ all waves polarized NE–SW with respect to those polarized NW–SE, or, to say it another way, to retard extraordinary rays in the NE and SW quadrants and ordinary rays in the NW and SE quadrants. If the crystal is optically positive (i.e., ordinary ray is fast), interference colors in the NE and SW quadrants advance 147 mμ while interference colors in the NW and SE retreat by an equal amount. An optically negative crystal will, of course, reverse quadrant colors, which subtract in the upper-right and lower-left and add in the upper-left and lower-right. The resulting colors are, again, dependent upon initial retardation, but first-order white adds 147 mμ to become first-order yellow and subtracts to become black. The black areas usually take the form of black dots immediately adjacent to the melatope, and stand out boldly against the white melatope and isogyres ($\Delta = 0$ mμ + 150 mμ). Positive crystals produce black dots in the upper-left and lower-right quadrants; negative crystals produce dots in the upper-right and lower-left.

The quartz wedge is useful for interference figures that have so many isochromes that the addition or subtraction of 147 mμ or even 550 mμ may go unnoticed. Since retardation of the quartz wedge is variable from zero to a high order, the color effects it produces are not static but appear as patterns of flowing color or migrating isochromes as the wedge is pushed slowly through the accessory slot, *thin end first* (Plate IV,C). The slow ray of the quartz wedge is again NE–SW, causing retardation of ordinary rays in the NW and SE and retardation of extraordinary rays in the NE and SW quadrants.

If the field of a uniaxial optic-axis figure is first-order white, addition of retardation causes the white to become first-order yellow, then orange, then red, and so on up the color scale. Subtraction changes first-order white first to black, then back to white, and thence up the scale.

An interference figure showing many isochromatic rings (Plate IV,C) lends itself well to this procedure because its isochromes move, either toward the melatope (addition of retardation) or away from the melatope (subtraction of retardation). Consider the series of very closely spaced isochromes from first-order black at the melatope to first-order red (i.e., black, white, yellow, orange, and red). Addition of increasing retardation by insertion of the quartz wedge causes the isochromes to move toward the melatope by changing first-order red to second-order blue, orange to red, yellow to orange, and white successively to yellow, orange, red and second-order blue. Subtraction causes the isochromes to move away from the melatope by changing white to black, yellow to white, orange to yellow, and red successively to orange, yellow, and white. Optically positive crystals, therefore, show isochromatic rings moving toward the melatope in the NE and SW quadrants and away from the melatope in the NW and SE quadrants as the quartz wedge is moved into the accessory opening. The reverse is of course true for optically negative crystals; again considering only the NE or upper-right quadrant, colors move toward the melatope for positive crystals and away from it for negative ones as the retardation of the quartz wedge increases.

OPTIC-AXIS FIGURE OFF-CENTER. Inclination of the optic axis will cause the melatope to be displaced from the center of the field; when inclination is great, the melatope will be beyond view,* and only one arm of the isogyre cross will be visible at any given time (Fig. 6-7,C). As the stage is rotated, isogyre arms sweep across the field, alternately parallel with the N–S and E–W cross hairs, and the melatope subscribes a circle beyond the field limits (Fig. 6-8). The melatope must lie on a visible isogyre beyond the limits of view. Consequently, if an E–W isogyre is observed to move downward (i.e., south) through the field and disappear, then the melatope is on an E–W line south of the field. If the following N–S isogyre moves to the left and disappears, the melatope is located in a SW direction beyond the field. Thus the visible field lies in the upper-right or NE quadrant, where addition of retardation means that the crystal is positive and subtraction means that it is optically negative (Plate V).

It is often possible to determine the optic sign of an off-center figure, but one can never be certain whether the crystal is uniaxial or biaxial unless a melatope can be seen.†

FLASH FIGURES. When the single optic axis of a uniaxial crystal is nearly horizontal, flash figures (Fig. 6-7,D) appear as a momentary darkening of the field with every 90° of stage rotation. Biaxial crystals in certain orientations yield a similar figure.

---

*An objective lens with high N.A. captures a larger cone of light than one with lower N.A. and hence, the melatope remains visible for greater inclination of the optic axis.

†Certain minerals that should be uniaxial (e.g., beryl, nepheline, and idocrase) often show a slight isogyre separation as the stage is rotated, indicating two optic axes. The mineral may be either biaxial with a very small $2V$ angle (see p. 125) or uniaxial with a strained crystal structure. The dilemma remains until additional data can be obtained for a specific mineral identification.

In the position of maximum field darkening, the darkened area is a very broad and diffuse cross with bright areas at the periphery of each quadrant (Plate VII,A). The optic axis is either N–S or E–W in the plane of the figure. Very slight stage rotation causes the darkness to leave the field, and careful observation reveals that two of the bright areas in alternate quadrants show increased retardation while the other two become dark. The darkened areas mark the position of the optic axis, and continued stage rotation places the optic axis in a 45° position where it connects the quadrants of lower retardation. The gypsum plate can now be used to determine whether the wave vibrating parallel to the optic axis (i.e., the extraordinary wave) is fast (negative) or slow (positive) (Plate VII,B).

The Search for Interference Figures

A student is often introduced to the subject of interference figures by being shown specially oriented crystal sections, where a predetermined figure is assured. In practice, he is faced with the problem of searching among many crystal grains or fragments to find one that will yield the most informative interference figure. The search for such a section or fragment can be very tedious and discouraging unless a systematic approach is used. The student is reminded that (1) the most useful uniaxial interference figure is a centered optic-axis figure, and (2) a centered optic-axis figure is produced by a grain or fragment that shows minimum birefringence. He will save time by scanning the slide with a medium-power objective lens and selecting a comparatively large grain or fragment that shows minimum birefringence. Such a grain should yield a useful figure. In thin section, all crystal grains are of equal thickness, and the one that shows the lowest interference color has the lowest birefringence. A crystal fragment, however, may show low-order color either because it is thin or because it has low birefringence. Thick fragments showing low-order color yield the best figures, and such fragments can often be recognized by their dispersion, which causes them to display pale and delicate shades of pink and blue instead of the normal first-order white or gray.

Index of Refraction

The Search for Principal Indices

A uniaxial crystal has a single index for all ordinary rays (n_ω) and a maximum or minimum index for extraordinary rays (n_ϵ). The principal indices of a given mineral are measurable values and thus constitute the most useful of all optical properties. They may be measured with great accuracy from crystal fragments but only crudely estimated in thin section.

The actual process of determining n_ω and n_ϵ is twofold: (1) fragments are examined with crossed nicols to select a particular fragment and to orient it to yield n_ω or n_ϵ, and (2) the index is measured by comparing the properly oriented fragment with its immersion medium under uncrossed nicols. The second phase of this process was described in Chapter 3 and will not be repeated here. This section is concerned only with the initial phase as it applies to uniaxial crystals.

A uniaxial crystal fragment under crossed nicols becomes dark when its inherent polarization directions parallel those of the nicols. In a position of extinction, the fragment passes only one ray, either ordinary or extraordinary (i.e., the N–S wave). In a position of nonextinction, the fragment passes components of two rays (hence retardation is possible), which form an intermediate index that is of little practical value. Reference to the uniaxial indicatrix shows that *fragments with minimum birefringence* (*optic axis centered*) *have a refractive index of n_ω in all positions of stage rotation; fragments showing maximum birefringence* (*flash figure*) *yield an index of n_ω in one extinction position and n_ϵ in the other*; and fragments showing intermediate birefringence yield an index of n_ω in one extinction position and some value between n_ω and n_ϵ in the other. The value of n_ω can be measured on any fragment, but the value of n_ϵ may be difficult to determine, as it is shown only by fragments that have maximum birefringence, which may be hard to recognize. The value of n_ϵ is either the maximum (positive crystal) or the minimum (negative crystal) that can be measured on any crystal fragment. If the student knows the optic sign, he also knows whether n_ω is greater or less than n_ϵ; if he does not know the optic sign, he can determine it by making this observation.

Color and Pleochroism

Any transparent substance is colored if it absorbs certain wavelengths of light and transmits others. Those minerals containing transition elements (e.g., Fe, Mn, Cr, etc.), especially those with double valence states, are usually colored, even in thin section or in small fragments. Allochromatic minerals are white or colorless when pure; when color is present, it results from minor impurities and is generally not apparent in mineral sections and fragments.

Ordinary and extraordinary rays may not only experience different velocities but different absorption effects with transmission through an anisotropic medium. Minerals that show a large difference in ray velocities (i.e., high birefringence) usually show large differences in absorption. Waves moving parallel to the optic axis (i.e., ordinary waves) all move with the same velocity, and all exhibit the same degree of absorption, because they all experience the same structural environment regardless of vibration direction. Those moving normal to the optic axis travel with maximum differences in velocity and exhibit maximum differences in absorption depending on their vibration directions. Waves vibrating normal to the optic axis (i.e., ordinary waves) travel at the same velocity, exhibit the same absorption, and experience the same environment as those moving parallel to the optic axis. Those vibrating parallel to the optic axis (i.e., extraordinary waves) experience maximum variance in structural environment, and thus exhibit maximum variance in velocity and absorption. Therefore, ordinary and extraordinary rays may be absorbed differently and thus produce different color effects; crystals that show this effect are said to be *pleochroic* or to display the property of pleochroism. Pleochroism in uniaxial crystals is twofold (dichroism), since the principal refractive indices are twofold.

Pleochroism is observed with *uncrossed nicols*, since it is a function of inherent color, not of interference color. As the microscope stage is rotated, a pleochroic crystal grain or fragment will change color if properly oriented. Basal sections show zero birefringence and zero pleochroism. Pleochroism is normally expressed as a formula, for example:

$$n_\omega = \text{pale yellow}, \quad n_\epsilon = \text{dark olive green}.$$

This implies that ordinary waves produce pale yellow and that extraordinary waves produce dark olive green. Unfortunately, quantitative measure of pleochroism is impractical; we therefore describe it as "extreme" to "weak." We have no way of predicting its magnitude, but strong pleochroism commonly accompanies high birefringence. Pleochroism may be "quantitative" (i.e., light and dark shades of the same basic color) or "qualitative" (i.e., different colors). Quantitative pleochroism is commonly expressed as $0 > E$ (or $E > 0$), to indicate that the ordinary ray is absorbed more than the extraordinary ray and that the crystal grain or fragment is darker when the optic axis lies in the E-W plane.

Crystal Orientation

Relationship Between Optical and Crystallographic Directions

For uniaxial crystals, the relationship between optical and crystallographic directions is invariant, since the unique optical direction (the optic axis) must coincide with the unique crystallographic direction (the *c*-axis). Cleavage planes and crystal faces represent other crystallographic directions that bear some systematic relation to the optic axis.

Extinction Angle

A simple quantitative measurement of this relationship is the extinction angle, or the angle between inherent vibration directions and an observed crystal face or cleavage plane (Fig. 6-9). Crystal faces and cleavages are limited to basal, prismatic, and pyramidal types in tetragonal, hexagonal, and trigonal crystals.

Basal pinacoids and basal cleavages define planes normal to the optic axis, and prisms and prismatic cleavages define planes parallel to the optic axis. Reference to the indicatrix shows that when the trace of any of these planes

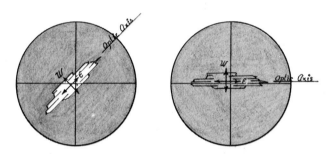

A - Parallel Extinction

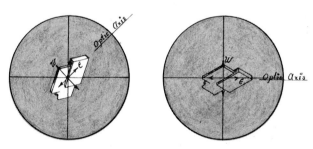

B - Symmetrical Extinction

FIGURE 6-9. Extinction of uniaxial crystals. Cleavage fragments of uniaxial crystals may become extinct (A) when the cleavage trace lies N–S and E–W (parallel extinction) or (B) when two intersecting cleavages are symmetrically bisected by the N–S and E–W directions (symmetrical extinction). Extinction is parallel to basal and prismatic cleavages and symmetrical to pyramidal cleavages (e.g., rhombohedral).

lies either N–S or E–W, the crystal must show extinction with crossed nicols, in which case the crystal is said to show *parallel extinction* (Fig. 6-9,A) to that cleavage or crystal face.*

Pyramidal crystal faces and cleavage planes are inclined to the optic axis and lie at some angle to the N–S and E–W cross hairs when the crystal is rotated to extinction between crossed nicols. Symmetry requires pyramidal planes to occur in symmetrically equivalent sets of three, four, six, eight, or twelve such that each plane shows an identical relationship to the c-axis.† Two or more of these planes or sets of planes should be visible in any section; the crystal extinguishes when the N–S and E–W cross hairs bisect the angle of intersecting pyramidal planes, showing symmetrical extinction (Fig. 6-9,B). The extinction angle, measured to one pyramidal plane, is equal and opposite to the extinction angle, measured to the other.

In summary, *uniaxial crystals show parallel extinction with respect to basal and prismatic planes and symmetrical extinction with respect to pyramidal planes.*

Sign of Elongation

Uniaxial crystal of linear habit or fragments produced by prismatic cleavages are elongated parallel to the c-axis. Inherent polarization planes lie parallel and perpendicular to the elongation, as shown by parallel extinction, and we may logically carry our observations one step further by determining whether the elongation parallels the plane of the fast or slow wave, which defines the *sign of the elongation*. This sign is determined by rotating the crystal section or fragment 45° from its extinction position so that its elongation lies NE–SW (Plate VIII). An accessory plate inserted into the microscope tube aligns the slow vibration direction of the accessory with crystal elongation, and the interference colors of the crystal either advance or retreat, in accord with the retardation of the accessory. If the slow wave of the accessory parallels the slow wave of the crystal, retardations add and the crystal is said to be *length-slow*, or to have *positive elongation*. Conversely, if slow wave parallels fast, retardations subtract and the crystal is said to be *length-fast*, or to have *negative elongation*. For elongated uniaxial crystals or fragments the extraordinary wave

---

*Elongated uniaxial crystals (e.g., tourmaline) must be elongated parallel to c, and hence show extinction parallel to elongation.

†Pyramidal cleavage is uncommon in uniaxial crystals except in the trigonal carbonates where rhombohedral, {1010}, cleavage is always present.

vibrates parallel to elongation, and the sign of elongation is the same as the optic sign. For crystal fragments showing several orders of retardation the quartz wedge is a most appropriate accessory; as the wedge is pushed into the accessory slot, the bands of color either move toward the thickest part of the fragment (indicating subtraction of retardation) or away from it (addition of retardation).

Pyramidal cleavage is uncommon in uniaxial crystals and does not produce elongation, but instead produces more-or-less equidimensional dipyramids or rhombs. Unless a crystal, crystal section, or crystal fragment shows some sort of crystallographic lineation or elongation, the sign of elongation is meaningless.

Cleavage of Uniaxial Crystals

Cleavage takes place along crystallographic planes in which bonding forces are weak and *cleavage must be identical in all symmetrically equivalent directions*.

Tetragonal, hexagonal, and trigonal crystals may show (1) no detectable cleavage at all, (2) basal cleavage, prismatic cleavage, or pyramidal cleavage in varying degrees of perfection, or (3) some combination of the cleavages in (2).

Basal Cleavage {001} or {0001} is defined by a single cleavage direction normal to the *c*-axis. It produces fragments that show no elongation and which lie with optic axes vertical. Fragments exhibit minimum birefringence and have a refractive index of n_ω and a centered optic-axis interference figure. In thin section, basal cleavage is apparent only in grains that show maximum birefringence, and produces elongation opposite in sign to the optic sign.

Prismatic cleavage is defined by two mutually perpendicular sets of cleavage planes parallel either to (100) and (010) or to (110) and (1$\bar{1}$0) for tetragonal crystals, and by three sets of cleavage planes intersecting at 60° parallel to the first-order {10$\bar{1}$0} or second-order {11$\bar{2}$0} prisms for hexagonal or trigonal crystals. All prismatic cleavages yield fragments elongated on *c* with parallel extinction and elongation similar in sign to the optic sign. Prismatic fragments or sections show maximum birefringence, flash interference figures, and an index of n_ϵ when elongated N–S and an index of n_ω when E–W. Basal sections show all prismatic cleavages, but only one cleavage is obvious in most other sections.

Pyramidal cleavage is defined by four sets of cleavage planes for tetragonal crystals and by three (rhombohedral) or six sets for trigonal and hexagonal crystals, all lying at an angle to *c*. Fragments show symmetrical extinction, no consistent elongation, intermediate birefringence, off-center interference figures, and an intermediate refractive index for the extraordinary ray. Thin sections show two pyramidal cleavages with symmetrical extinction.

CHAPTER 7

Optical Crystallography of Biaxial Crystals

Biaxial Anisotropism

Crystals are anisotropic when their physical properties, notably light propagation, differ with direction. Orthorhombic, monoclinic, and triclinic crystals are defined by three unique crystallographic directions, a, b, and c, and their crystal structure causes the propagation of light to differ in all three dimensions. A ray of light moving in any direction within a biaxial crystal experiences a different environment from that of any other ray, except as rays are related by optical symmetry. Crystals of these systems are anisotropic and said to be biaxial because two optic axes replace the single optic axis of uniaxial crystals.

The Biaxial Indicatrix

It is possible to derive a biaxial indicatrix by the same procedures used previously to arrive at its uniaxial analog—that is, by making use of simple observations to develop a ray velocity surface for a biaxial crystal and ultimately its equivalent indicatrix. But because such a procedure would be more involved, for a biaxial crystal, and because the student should by now be at ease with the notion of using an imaginary geometrical surface to represent index of refraction and vibration directions of light waves within an anisotropic crystal, let us merely present the biaxial indicatrix and derive from it, largely as a practical exercise in its use, the biaxial ray velocity surfaces.

Optical Directions

Three unique crystallographic directions (a, b, c) require three dimensions of unique crystal structure and, logically, *three unique optical directions*, designated *X, Y,* and *Z* (Fig. 7-1). Optical directions (*X, Y, Z*) are always mutually perpendicular, in contrast to crystallographic directions (*a, b, c*), which are not mutually perpendicular in monoclinic and triclinic crystals. Optical and crystallographic directions cannot always coincide. Planes *XY, XZ,* and *YZ* are known as *principal sections* of the indicatrix.

Principal Indices of Refraction

Construction of the uniaxial indicatrix was based upon *two principal indices*: n_ω, the index for the ordinary wave, and n_ϵ, the index for the extraordinary wave, which vibrates parallel to the unique optical direction. Construction of the biaxial indicatrix is based upon *three principal indices* of refraction: n_α, n_β, and n_γ (Fig. 7-1). In a biaxial medium, no obvious distinction exists between ordinary and extraordinary rays, and essentially all rays are extraordinary.* Thus n_α *is the refractive index of the crystal for waves vibrating parallel to the unique optical direction X*, n_β *is the index for waves vibrating parallel to Y, and n_γ is the index for waves vibrating parallel to Z.*

Although either principal index of a uniaxial crystal may be the larger, the principal biaxial indices are defined with n_α as the smallest, n_γ as the largest, and n_β as the intermediate.

Triaxial Ellipsoid

The biaxial indicatrix is a triaxial ellipsoid formed by three principal axes of length $2n_\alpha$, $2n_\beta$, and $2n_\gamma$ which lie along the *X, Y,* and *Z* optical directions, respectively (Fig. 7-2). The triaxial ellipsoid is represented by the equation

$$\frac{x^2}{n_\alpha^2} + \frac{y^2}{n_\beta^2} + \frac{z^2}{n_\gamma^2} = 1.$$

---

*All waves moving within a principal section and vibrating normal to it travel with equal velocity and may be considered ordinary.

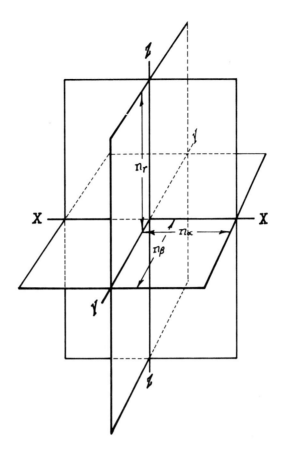

FIGURE 7-1. Optical directions and principal indices of refraction. Optical directions X, Y, and Z are mutually perpendicular, forming principal planes XY, XZ, and YZ. The principal refractive indices n_α, n_β, and n_γ are the refractive indices for light waves vibrating parallel to the optical directions X, Y, and Z, respectively. Relative magnitude is always $n_\gamma > n_\beta > n_\alpha$.

The ellipsoid has only three planes of symmetry formed by principal sections on the XY-, XZ-, and YZ-planes.

With two very important exceptions, sections through the center of the triaxial ellipsoid are ellipses. Sections containing Y are ellipses with one semi-axis equal to n_β. With rotation about Y, the other semi-axis ranges from n_α to n_γ; at some angle, the semi-axis must equal n_β, and at that angle the section becomes *circular*. A second circular section exists at a similar, but opposite, angle, as required by symmetry.

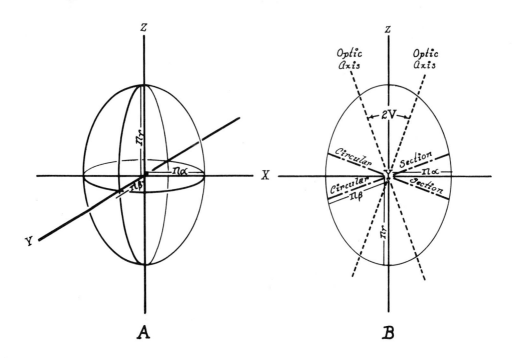

FIGURE 7-2. Biaxial indicatrix. (A) The triaxial ellipsoid is based upon three principal axes of length $2n_\alpha$, $2n_\beta$, and $2n_\gamma$ measured along the optical directions X, Y, and Z, respectively. All sections of the ellipsoid are elliptical except two circular sections of radius n_β. (B) The optic plane is the XZ-plane, Y is normal to the XZ-section, and the two circular sections of radius n_β are also normal to XZ and intersect on Y. Perpendicular to each circular section is an optic axis. The optic axes intersect at acute angle $2V$ and define the optic plane.

The Optic Axes

The two circular sections of a biaxial indicatrix carry the same implications as the single circular section of the uniaxial indicatrix. The direction perpendicular to each circular section is an *optic axis*, along which all light rays advance with equal velocity, and waves are not confined to vibration in particular directions. The plane of the two optic axes is the XZ-plane, designated the *optic plane*; the Y-direction, normal to the optic plane, is called the *optic normal*.

*The 2V Angle**

The acute angle of intersection between optic axes is the 2V angle; just as isotropic crystals become a special case of uniaxial crystals where $n_\epsilon = n_\omega$, so are uniaxial crystals a special case of biaxial crystals where $2V = 0°$.

Optical directions X and Z bisect the axial angles. The direction (X or Z) that bisects the acute angle (i.e., the 2V) is called the *acute bisectrix* (Bxa), and the direction (Z or X) that bisects the obtuse angle is the *obtuse bisectrix* (Bxo).

The notation "2V" implies that one-half the acute angle is V, represented as V_X when measured from X and as V_Z when measured from Z.

The Optic Sign

If n_β were decreased until it equalled n_α, the circular sections would merge to a single circle of radius n_α in the XY-plane, the optic axes would merge to a single axis along Z, and the biaxial indicatrix would become the indicatrix of uniaxial positive crystal (i.e., a prolate spheroid) (Fig. 7-3). Increasing n_β again causes circular sections and optic axes to separate with Z as acute bisectrix until the 2V (i.e., $2V_Z$) angle becomes 90°. Further increase in n_β causes X to become the acute bisectrix, and the 2V (i.e., $2V_X$) angle decreases until n_β equals n_γ and the indicatrix becomes the indicatrix of a uniaxial negative crystal (i.e., an oblate spheroid).

When Z is the acute bisectrix, the indicatrix is defined as optically positive; when X is the acute bisectrix, the indicatrix is defined as negative. From the above, it should be apparent that the indicatrix is positive when the value of the n_β index is nearer n_α than n_γ and negative when n_β is nearer n_γ.†

When 2V equals 90°, the above definition no longer holds, and both the indicatrix and the crystal it represents are neither optically positive nor negative but optically neutral. The closer 2V approaches 90°, the greater the practical difficulties encountered in determining optic sign.

Crystallographic Orientation of the Biaxial Indicatrix

Like its uniaxial counterpart, a biaxial indicatrix must be assigned a specific crystallographic orientation to be of any use. As a generalization, we may note

---

*The term "axial angle" is often used to refer to the angle, either obtuse or acute, between optic axes.

†Caution is advised when 2V closely approaches 90°, as this general statement may not hold true.

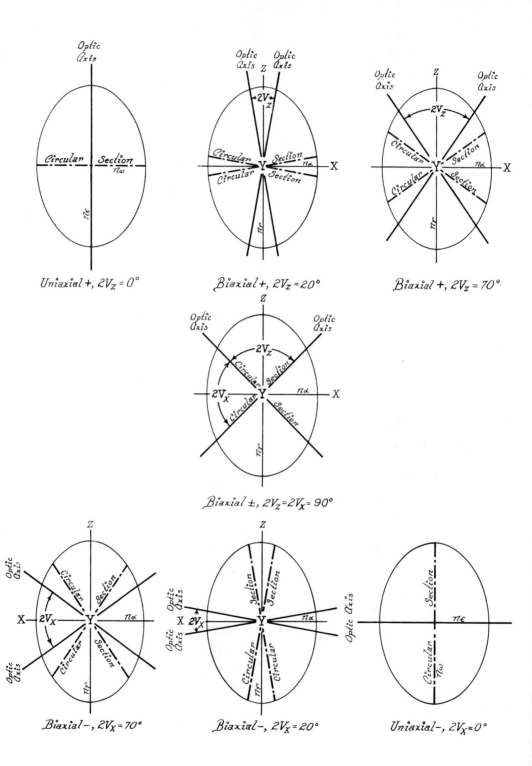

FIGURE 7-3. Relation of uniaxial and biaxial indicatrixes. Uniaxial positive is a special case of biaxial positive, where $2V_Z = 0°$ and $n_\alpha = n_\beta(n_\omega) < n_\gamma(n_\epsilon)$. The indicatrix is positive so long as Z bisects the acute $2V$ angle; when $2V_Z = 90°$, so does $2V_X$ and the indicatrix is neutral; as X becomes the acute bisectrix, the indicatrix becomes biaxial negative, and when $2V_X = 0°$ and $n_\gamma = n_\beta(n_\omega) > n_\alpha(n_\epsilon)$ the indicatrix is uniaxial negative.

that symmetry planes of the indicatrix (i.e., principal sections) must parallel crystallographic symmetry planes, so far as the latter exist.

Orthorhombic crystals show a maximum symmetry of three mutually perpendicular symmetry planes defined by three mutually perpendicular crystallographic directions *a*, *b*, and *c*. The biaxial indicatrix shows orthorhombic symmetry, which must parallel that of the crystal. Principal sections parallel crystallographic symmetry planes, and optical directions parallel unique crystallographic directions (Fig. 7-4,A). To fix the indicatrix for a given orthorhombic mineral, we need only determine which of the optical directions (*X*, *Y*, or *Z*) parallels each crystallographic direction (*a*, *b*, and *c*).

Monoclinic crystals possess no more than one symmetry plane {010} and one symmetry axis (*b*-crystallographic axis). One principal section of the indicatrix must parallel the single symmetry plane, causing one optical direction, *X*, *Y*, or *Z*, to coincide with the *b*-crystallographic axis (Fig. 7-4,B). To fix the indicatrix for a given monoclinic crystal, one must determine which optical direction parallels *b* and establish an angular relationship between one other optical direction and one other crystallographic axis.

Triclinic crystals show no elements of symmetry other than a center of symmetry, and one can make no predictions about crystallographic and optical correlation. For a given triclinic mineral, one must establish an angular relationship between each crystallographic direction and its nearest optical direction (Fig. 7-4,C).

Use of the Biaxial Indicatrix

The biaxial indicatrix is constructed on the basis of the same principles as its uniaxial analog; it serves the same function and yields information about vibration directions and refractive indices in the same way. An appropriately constructed and properly oriented indicatrix is imagined to exist within the crystal section or fragment such that the observer's line of sight always passes through the indicatrix center. A section through the center of the indicatrix normal to the direction of light propagation produces an ellipse or a circle. For light rays moving parallel to *X*, *Y*, or *Z*, the elliptical section thus formed

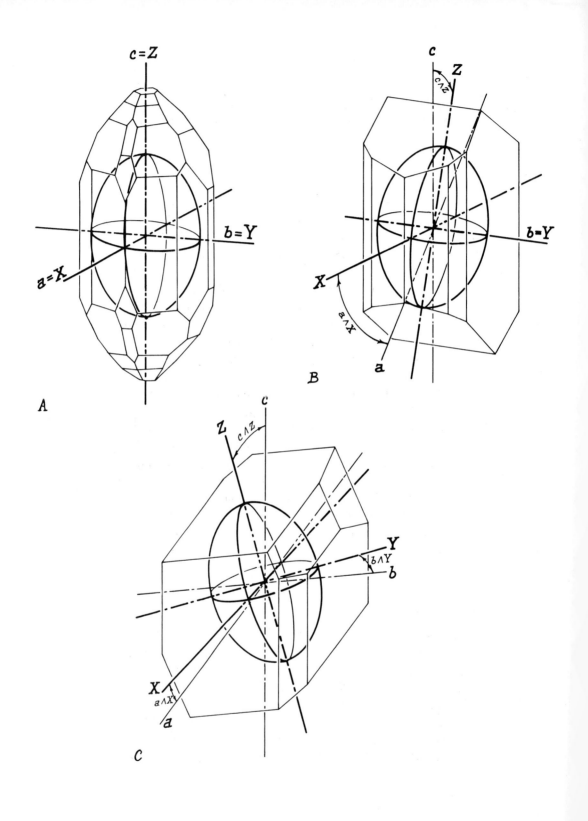

FIGURE 7-4. Optical orientation. (A) Orthorhombic Crystals. Optical directions X, Y, and Z must parallel crystallographic directions a, b, and c in one of six possible orientations. (B) Monoclinic Crystals. One optical direction X, Y, or Z must parallel the b-crystallographic direction. (C) Triclinic Crystals. There are no fixed relationships between optical and crystallographic directions.

indicates by the length of its semi-axes the indices of refraction for the two mutually perpendicular waves, each of which vibrates parallel to the axis representing its index (Fig. 7-5,A,B,C). For all other directions of light propagation, the perpendicular section represents only approximately the vibration directions and refractive indices of the two mutually perpendicular advancing waves (Fig. 7-5,D,E). Such an approximation is sufficient for most purposes, but a more accurate representation of vibration directions and indices for a random ray direction is given by an indicatrix section parallel to the plane tangent to the indicatrix surface at its intersection with the random ray direction.

Biaxial Ray Velocity Surfaces

Although of little practical value, biaxial ray velocity surfaces may be derived as an exercise in the use of the biaxial indicatrix.

For clarity, let us construct only one octant of the ray velocity surfaces as defined by X, Y, and Z of Figure 7-6.

Imagine a point light source within a crystal represented by the indicatrix in Figure 7-5. From the origin of X, Y, and Z, rays and their associated waves advance outward in all directions and at a given instant are arrested to form appropriate ray velocity surfaces.

The indicatrix tells us that along X, two mutually perpendicular waves advance with different velocities. One wave vibrates parallel to Z and advances with a velocity V_γ, inversely proportional to n_γ, and the other wave vibrates parallel to Y and advances with a greater velocity V_β, inversely proportional to n_β. Along Y, the faster wave moves with velocity V_α and vibrates parallel to X; the slower wave advances with velocity V_γ and vibrates parallel to Z; along Z, the faster ray advances with velocity V_α and vibrates parallel to X, and the slower ray advances with velocity V_β and vibrates parallel to Y (Fig. 7-6,A).

Similar reasoning, with help of the indicatrix, shows that along any propagation direction within the XY-plane two mutually perpendicular waves advance from the origin (Fig. 7-6,B). The slower wave always vibrates parallel

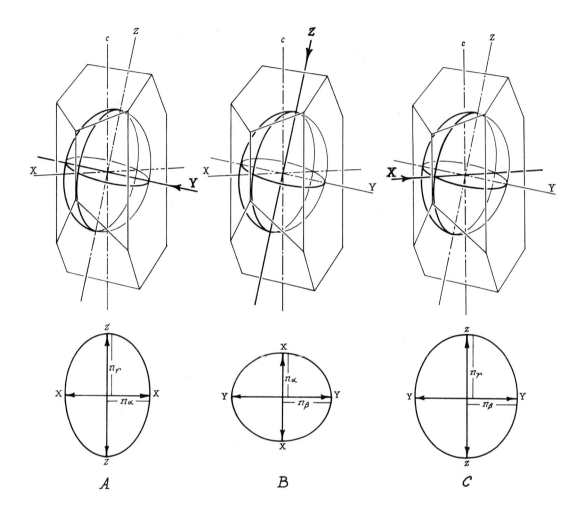

FIGURE 7-5. Use of the indicatrix. (A) Unpolarized light strikes the crystal parallel to Y. The indicatrix section perpendicular to Y indicates that two mutually perpendicular waves pass through the crystal, one vibrating parallel to Z with a refractive index of n_γ and the other vibrating parallel to X with refractive index n_α. (B) Unpolarized light strikes the crystal parallel to Z. The indicatrix section perpendicular to Z shows the crystal to pass one wave vibrating parallel to Y with refractive index n_β and another vibrating parallel to X with refractive index n_α. (C) Unpolarized light strikes the crystal parallel to X. The indicatrix section normal to X shows the crystal to pass only light waves vibrating parallel to Z

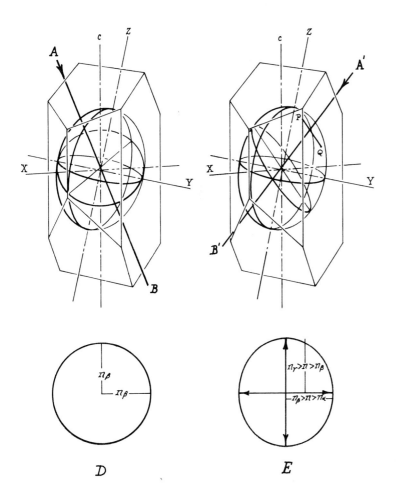

(refractive index n_γ) and light waves vibrating parallel to Y (refractive index n_β). (D) Light moving from A to B strikes the crystal normal to a circular section, indicating the crystal does not restrict vibration directions and that all waves experience the same refractive index n_β. (E) Light waves moving from A' to B' strike the crystal in a random direction. Vibration directions and refractive indices are best represented by the indicatrix section parallel to PQ and tangent at the point of contact. The student is seldom concerned with random ray directions.

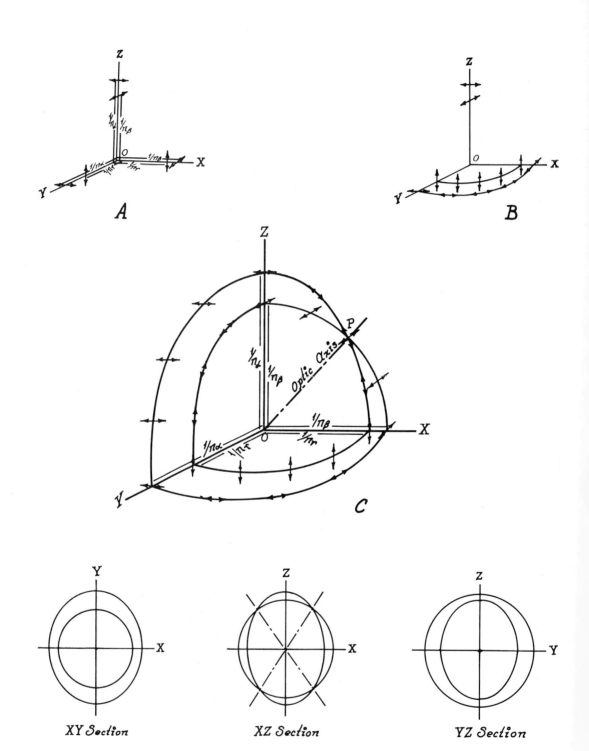

FIGURE 7-6. Biaxial ray velocity surfaces. Within a biaxial crystal, light moves outward in all directions from point O. Consider the advance of all waves after time interval Δt. (A) Along Z, the wave vibrating parallel to X has progressed a distance $V_\alpha \Delta t \propto 1/n_\alpha$, and the wave vibrating parallel to Y has advanced $V_\beta \Delta t \propto 1/n_\beta$. Along Y, the wave vibrating parallel to X has advanced $V_\alpha \Delta t \propto 1/n_\alpha$, and the wave vibrating parallel to Z has moved $V_\gamma \Delta t \propto 1/n_\gamma$. Along X, the wave vibrating parallel to Y has progressed the distance $V_\beta \Delta t \propto 1/n_\beta$ and the wave vibrating parallel to Z has advanced $V_\gamma \Delta t \propto 1/n_\gamma$. (B) All waves moving outward from O in the XY-plane either vibrate parallel to Z and move with velocity $V_\gamma \propto 1/n_\gamma$ or vibrate in the XY-plane tangent to the indicatrix surface and advance with velocities ranging from $V_\alpha \propto 1/n_\alpha$, along Y, to $V_\beta \propto 1/n_\beta$, along X. (C) Similar reasoning about all principal planes can be used to construct one octant of the biaxial ray velocity surface, which is a double surface that intersects only at point P in the XZ-plane. All light waves along OP (an optic axis) advance with equal velocity $V_\beta \propto 1/n_\beta$ and are not confined to vibrate in any particular plane. Principal sections of the ray velocity surfaces are shown at the bottom of the figure.

to Z and travels with velocity V_γ, forming a circular section in the XY-plane. (These may be considered ordinary waves.) The faster ray moves with velocity ranging from V_β, for a ray along X, to V_α, for a ray along Y. These rays form an elliptical section in the XY-plane, and each associated wave vibrates in the plane tangent to the section. Within the YZ-plane, the faster waves vibrate parallel to X and advance with velocity V_α, forming a circular section with YZ; the slower waves range from V_γ, along Y, to V_β, along Z, forming an elliptical section with YZ and vibrating tangent to the section. Within the XZ-plane, one set of waves vibrating parallel to Y advances with velocity V_β and forms a circular section with XZ; the other set, vibrating in the XZ-plane, advances with velocity ranging from V_α, along Z, to V_γ, along X, and forms an elliptical section with XZ.

At one point P (Fig. 7-6,C) in the XZ-plane the elliptical and circular sections cross, indicating that all rays advancing from the point of origin to that point move with equal velocity. Thus the direction of an optic axis is defined.*

Three-dimensional surfaces may be formed by connecting XY, YZ, and XZ sections to yield two odd-shaped ray velocity surfaces that are tangent at the four points where optic axes emerge.

---

*These are called secondary optic axes in contrast to the primary optic axes of the indicatrix. A more complex indicatrix theory would suggest that primary and secondary optic axes do not quite coincide, though this is seldom of concern to the beginning student.

CHAPTER 8

Biaxial Crystals and the Petrographic Microscope

Interference Colors and Birefringence

The section on interference colors (Chapter 6, pp. 89 to 101) needs neither repetition nor modification here, since principles of retardation, birefringence, and interference apply equally well to uniaxial and biaxial crystals. Birefringence of a biaxial crystal changes with direction from zero to a maximum value, the latter being characteristic of the mineral. For rays traveling parallel to an optic axis, the crystal shows zero birefringence, and only for rays moving parallel to the optic normal does the crystal show maximum birefringence ($n_\gamma - n_\alpha$). Light rays moving along X experience an intermediate birefringence ($n_\gamma - n_\beta$), and rays advancing parallel to Z a birefringence of $n_\beta - n_\alpha$.

The principles used to establish the relationship between birefringence and crystal structure for uniaxial crystals (p. 76) apply also to biaxial crystals, but the orientation of highly polarized linear or planar ions is more difficult to predict. Crystals containing such ions (e.g., carbonates and nitrates) have the potential for large birefringence, and crystals containing highly symmetrical ions (e.g., sulfates and phosphates) tend to have weak birefringence. In biaxial crystals, polarized units have many possible arrangements; asymmetrical units may be present in low-birefringent crystals, and highly symmetrical units may be present in crystals with high birefringence. The highly symmetrical silica tetrahedron occurs in symmetrical distribution in the framework structures of quartz, feldspar, feldspathoids, and zeolites; crystals of these minerals all show low birefringence. Silica chain and sheet structures yield tetrahedrons in linear and planar distribution and favor crystals of greater birefringence (e.g., pyroxenes, amphiboles, micas, talc).

Biaxial Interference Figures

Isochromatic Color Bands (Isochromes)

A biaxial interference figure is a stable pattern of isochromes, resulting from wave retardation, with superimposed isogyres that shift position with stage rotation and result from coincidence of wave and microscope vibration directions. To understand the formation of isochromes, consider a fragment or section of a biaxial positive crystal lying in the convergent light of conoscopic illumination between crossed nicols, with its acute bisectrix (Z) vertical (Fig. 8-1). Two rays travelling in the XZ-plane and inclined at an angle V to the Z-direction will pass along the optic axes, experience zero birefringence, and form two separate melatopes on the trace of the optic plane. All other rays undergo retardation, which increases with inclination to the optic axes, and produce equal-retardation, or isochromatic, bands that encircle the melatopes. Between optic axes in the XZ-plane, rays show greater birefringence as they approach the vertical, reaching a value of $(n_\beta - n_\alpha)$ along Z. From the melatopes toward the center of the field, birefringence increases but thickness slowly decreases; in all other directions from the melatope, both birefringence and thickness increase. Retardation, therefore, increases in all directions from the melatopes but at a slower rate toward the center of the field causing isochromes to be egg-shaped about each melatope (Fig. 8-1,A).

Isogyres

Conoscopic rays from the condensing lens converge to a focal spot in the crystal and diverge upward to the objective lens. The simplest way to understand how the isogyres of a biaxial interference figure are formed is to imagine that an appropriate indicatrix is centered at the focal spot and then to construct a series of lines on the surface of the indicatrix to represent vibration directions of all these diverging rays. Figure 8-2 shows an indicatrix marked so that each intersection on the surface of the indicatrix represents the vibration directions of two mutually perpendicular waves moving along a ray from the center of the indicatrix to the point of intersection, as determined by normal indicatrix procedures (Fig. 7-5). Any portion of this pattern projected orthographically (see p. 180) onto a plane is called a *skiodrome*.

With the acute bisectrix at the center of the figure, lines representing vibration directions appear as shown in Figure 8-3, and isogyres appear where vibration directions are N–S and E–W (see also Fig. 8-4). When the optic plane is 45° to the N–S and E–W cross hairs, isogyres appear as hyperbolas with melatopes at points of maximum curvature, and when the optic plane parallels a cross hair, isogyres join at the center of the field to form a cross, often resembling that of a uniaxial optic-axis figure. Further stage rotation causes the isogyres to separate, join, and again separate in opposite quadrants. Melatopes always remain equally spaced on the trace of the optic plane, which rotates with the stage, as do the optic normal and all isochromes. If isogyres separate and completely leave the visible field in the 45° position, melatopes lie outside the field and subscribe a circle beyond the limits of view as the stage is rotated.

The 2V Angle and Its Measurement

THE 2E ANGLE. Because diverging light rays are refracted as they leave a crystal (Fig. 8-5), the observed angle between optic-axis rays is greater than $2V$ and is known as the *2E angle*. From Snell's Law, V and E are related by

$$\frac{\sin E}{\sin V} = \frac{n_\beta}{1} \quad \text{or} \quad \sin E = n_\beta \sin V,$$

where n_β is the refractive index for rays passing along an optic axis.

MALLARD'S METHOD (*Fig. 8-5*). Separation of the melatopes* and maximum separation of the isogyres in the 45° position are functions of the $2E$ angle and of the optical elements of the microscope, such that

$$D = K \sin E \quad \text{or} \quad D = K n_\beta \sin V,$$

where D is one-half the observed melatope separation and K is a constant (Mallard's Constant) that depends upon the lens elements of a particular microscope. A linear or grid micrometer scale in the ocular is used to measure

---

*Melatope separation is independent of crystal thickness.

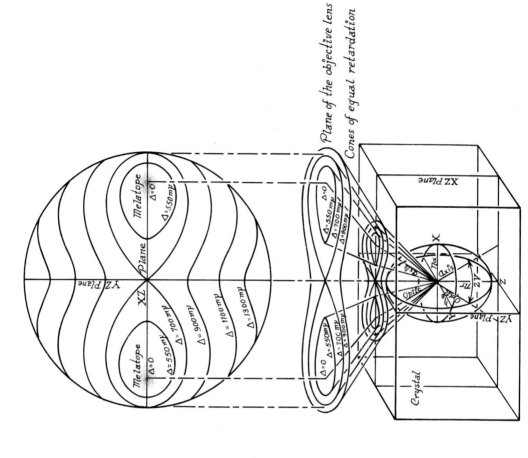

FIGURE 8-1. Formation of isochromes. An isochrome is a band of equal retardation, and hence uniform color. Light rays passing along optic axes experience zero retardation and form melatopes. From the melatopes, retardation increases in all directions to a maximum for light rays passing along the optic normal Y. (A) Cones of equal retardation form egg-shaped isochromes about the melatopes in the interference figure. (B) Retardation in the optic (XZ) plane increases slowly from the melatopes toward the center of the field because the length of the light path decreases as t as birefringence increases to $(n_\beta − n_\alpha)$. For light paths more inclined than the optic axes, both thickness (i.e., the length of the light path in the crystal) and birefringence increase, and retardation increases rapidly. (C) Retardation in the YZ-plane increases from a minimum $t(n_\beta − n_\alpha)$ at the center of the field. Both thickness and birefringence increase with inclination of the rays.

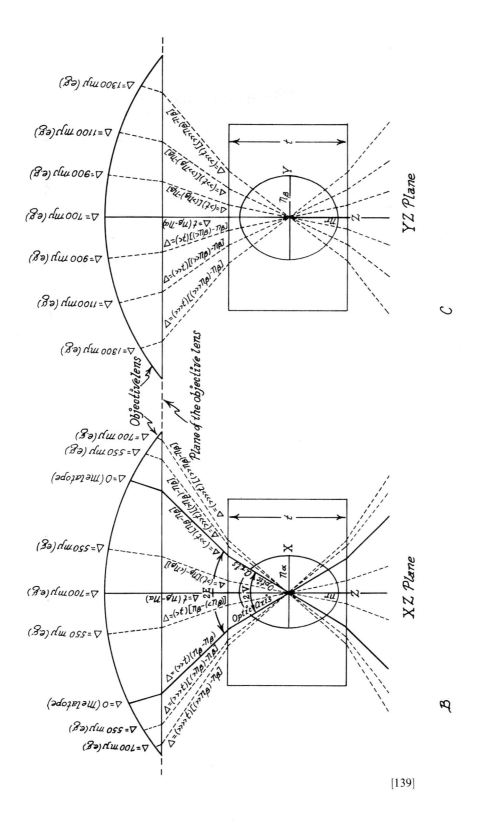

[139]

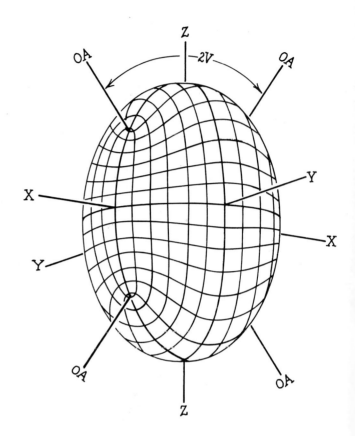

FIGURE 8-2. Biaxial indicatrix marked with vibration directions. Two sets of intersecting ellipses drawn on the reference surface of an indicatrix form sets of intersecting lines which, at every point on the surface, intersect at right angles. Each intersection represents mutually perpendicular inherent vibration directions for the waves that pass through that point. Ellipses represent the summation of vibration directions for all possible directions of propagation as derived from normal indicatrix procedures.

D, and a crystal of known $2V$ is used to determine K for given optical elements. The units of D and K are unimportant. Under favorable conditions, $2V$ may be measured to within one or two degrees.

For a large optic angle, more than about 60°, the melatopes do not commonly lie within the visible field. An oil-immersion objective captures a large cone of divergent light and may bring the melatopes into the visible field, in which case

$$D = K \frac{n_\beta}{n_{\text{oil}}} \sin V.$$

MICHEL-LEVY METHOD (*Fig. 8-6*). In 90° of stage rotation, biaxial isogyres must move in from points of maximum separation in opposite quadrants, join at the center, and retreat to maximum separation in the other two opposing

FIGURE 8-3. Formation of isogyres with stage rotation—acute bisectrix. Isogyres are formed by extinction wherever vibration directions are N–S and E–W. Stage rotation causes two separate isogyres to join at the field center when the optic (XZ) plane is N–S or E–W and to separate into opposite quadrants when the optic plane is NE–SW or NW–SE. Although isogyres join and separate, melatopes retain constant separation as determined by the 2V angle.

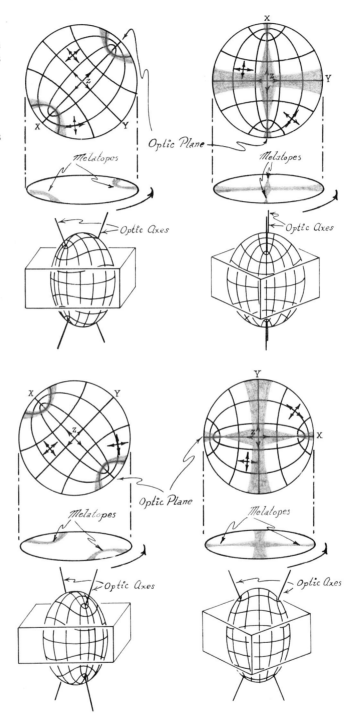

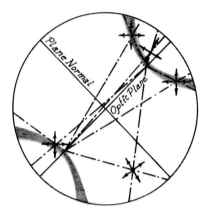

FIGURE 8-4. Geometrical construction of vibration directions. A simple geometrical construction may be used to indicate vibration directions for the interference figure proper. From each melatope, a line is constructed through any selected point in the figure, and the directions that bisect the angles formed by these two intersecting lines represent vibration directions in the plane of the figure at that point.

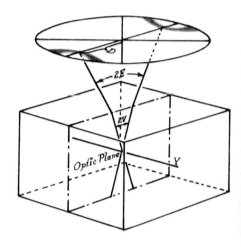

FIGURE 8-5. The $2E$ angle and Mallard's construction. Upon leaving the crystal plate, light rays moving along optic axes separate to an angle $2E$, through refraction at the interface, where $\sin E = n_\beta \sin V$. Melatope separation equals $2D$ where $D = K \sin E$ or $K n_\beta \sin E$; K is Mallard's constant.

quadrants. For large values of $2V$, maximum separation is greater, and the isogyres must move more quickly than for small values of $2V$. From their position of contact at the center of the visible field, isogyres can, by stage rotation, be made to separate a distance equal to the arbitrary diameter of a circle etched on a special glass disk inserted in the ocular. The number of degrees of stage rotation necessary to cause this separation is proportional to $2V$. This procedure yields very poor accuracy when the bisectrix is slightly inclined, but it applies for both acute and obtuse bisectrix figures and perhaps affords the best method of distinguishing the two (see p. 149).

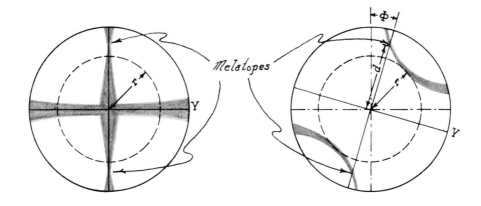

FIGURE 8-6. Michel-Levy method of measuring $2V$—separation of isogyres with stage rotation. The angle of stage rotation (Φ) necessary to cause an isogyre separation equal to some measurable distance r, where $r < d$, is a function of the angle between optic axes (i.e., $2V$ or $180° - 2V$). In practice, one may use the periphery of the visible field as the reference circle of radius r. If isogyres never separate far enough to leave the visible field in 45° of stage rotation (i.e., if $d < r$), the bisectrix figure is acute (Bxa) and $2V$ is less than 70°, assuming that $n_\beta > 1.50$ and that the N.A. of the objective lens is about 0.85. If isogyres leave the visible field within less than 15° of stage rotation ($\Phi < 15°$), the bisectrix figure is almost certainly obtuse (Bxo) and the $2V$ angle is again less than about 70°. If the isogyres leave the field in more than 15° of stage rotation, the $2V$ angle is "large."

VISUAL ESTIMATE FROM ACUTE BISECTRIX FIGURE (*Fig. 8-7*). From the relation

$$\text{N.A.} = n \sin u = \sin E_{\max},$$

where $E_{\max}$ is the largest visible E (see p. 33) we obtain

$$\sin V_{\max} = \frac{\text{N.A.}}{n_\beta}.$$

For a given objective lens (N.A.) and a given n_β, there is a value of V at which the melatopes lie on the edge of the visible field, and the isogyres retreat just to the field edge in the 45° position.

N.A.	n_β	Max. Visible 2V
0.85	1.40	75°
	1.50	69°
	1.60	64°
	1.70	60°
	1.80	56°

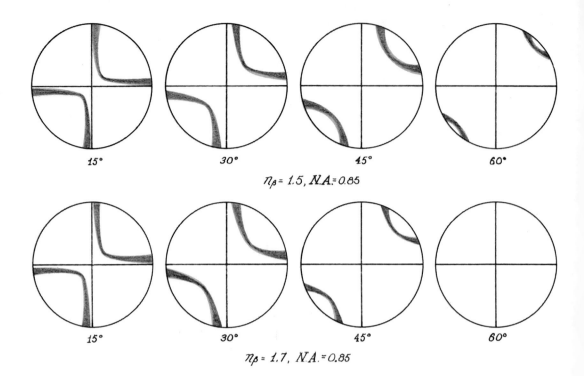

FIGURE 8-7. Visual estimate of 2V angle from acute bisectrix figure. Mallard's equation ($D = Kn_\beta \sin V$) reveals that melatope separation (i.e., maximum separation of isogyres in a 45° position) for a given 2V angle is a function of the refractive index n_β and of a microscope constant K, which depends upon the numerical aperture of the objective lens.

Since n_β for most common minerals lies between 1.5 and 1.7, students can usually learn to estimate 2V visually from maximum isogyre separation by observing figures of known 2V. Although accuracy is poor (say ±15°), the results are usually worth the effort.

VISUAL ESTIMATE FROM OPTIC-AXIS FIGURE* (*Fig. 8-8*). When an optic axis is vertical, the figure is a single hyperbolic isogyre with a single melatope at the point of maximum curvature at the center of the field. In the 45° position, the isogyre shows maximum curvature, and one can learn to estimate the 2V angle within ±10° by reference to Figure 8-8. Note that the isogyre is straight when 2V = 90°, bends at a right angle when 2V = 0° (i.e., produces a uniaxial cross), and lies about halfway between when 2V = 45°.

---

*This procedure is probably the most useful for estimating 2V angles, and the student is encouraged to use it often.

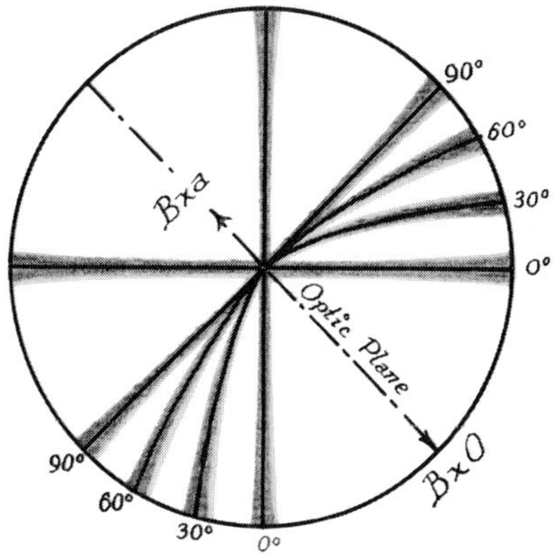

FIGURE 8-8. Visual estimate of 2V angle from an optic-axis figure. Maximum curvature of the single isogyre of an optic-axis figure ranges continuously from no curvature, when 2V = 90°, to a right-angle bend (i.e., the uniaxial cross), when 2V = 0°, and curvature is always convex toward the acute bisectrix.

GEOMETRICAL METHOD (*Fig. 8-9*). The 2V angle is related to principal indices of refraction by the expressions

$$\cos^2 V_z = \frac{n_\alpha^2 (n_\gamma^2 - n_\beta^2)}{n_\beta^2 (n_\gamma^2 - n_\alpha^2)} \quad \text{and} \quad \cos^2 V_x = \frac{n_\gamma^2 (n_\beta^2 - n_\alpha^2)}{n_\beta^2 (n_\gamma^2 - n_\alpha^2)}$$

or, less accurate but more practical:

$$\cos^2 V_z \simeq \frac{n_\gamma - n_\beta}{n_\gamma - n_\alpha} \quad \text{and} \quad \cos^2 V_x \simeq \frac{n_\beta - n_\alpha}{n_\gamma - n_\alpha}.$$

Refraction indices n_α, n_β, n_γ and 2V are not all independent variables; if any three are known, the fourth can be determined. For convenience, these relationships are expressed graphically in Figure 8-9.

Types of Interference Figures and Their Optic Signs

ACUTE BISECTRIX (Bxa) FIGURE. An acute bisectrix interference figure results when the acute bisectrix, Z or X, is vertical.

The problem of determining optic sign reduces to one of discovering the relative velocities of the two mutually perpendicular waves moving along the acute bisectrix (Fig. 8-10). The two waves emerge at the center of the figure; one wave vibrates parallel to the optic normal (Y) and advances with velocity $V_\beta \propto 1/n_\beta$; the other vibrates parallel to the optic plane and advances with velocity $V_\alpha \propto 1/n_\alpha$ when Z is the acute bisectrix and the crystal is optically positive, or with velocity $V_\gamma \propto 1/n_\gamma$ when X is the acute bisectrix and the crystal

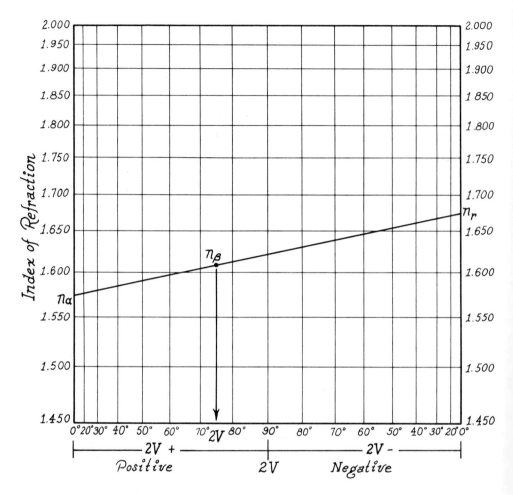

FIGURE 8-9. Graphical representation of n_α, n_β, n_γ, and $2V$. [After Mertie, 1958.] Knowing any three of the variables n_α, n_β, n_γ, or $2V$ fixes the fourth. As an example, $n_\alpha = 1.575$, $n_\beta = 1.610$, and $n_\gamma = 1.675$; thus $2V$ is 75°, and the crystal is optically positive.

is optically negative. The obvious solution is to orient the optic plane in the 45° position parallel to the slow vibration direction of the accessory planes, and any accessory should then show which is the faster ray (Plate IX).

For a more complete understanding, consider the acute bisectrix figure of a *positive* crystal (Fig. 8-10). Every point on the figure denotes the emergent path of two mutually perpendicular waves (Fig. 8-3). For all points along the trace of the optic normal (Y), one wave vibrates parallel to X, where the refractive

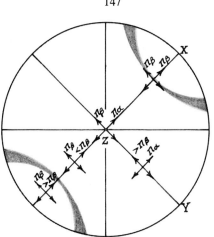

FIGURE 8-10. Acute bisectrix figure of a biaxial positive crystal showing refractive indices for selected waves. Those waves vibrating parallel to principal optical directions X and Y vibrate in the plane of the page and experience principal refractive index n_α and n_β, respectively. Other vibration directions are inclined to the page, and their waves experience intermediate indices of refraction.

index is n_α, and its companion wave vibrates in the ZY-plane, where the refractive index is between n_β and n_γ. The second is, therefore, always the slower wave. For all points along the trace of the optic plane, one wave vibrates parallel to Y, where the refractive index is n_β, and its normal companion wave vibrates in the XZ-plane, where the refractive index ranges from n_α for rays along Z to n_γ for rays along X. The second wave may, therefore, be either faster or slower. For rays between Z and an optic axis, the wave in the XZ-plane has an index between n_α along Z and an index of n_β along the optic axis, and is the faster wave. For rays between X and the optic axis, the wave in the XZ-plane has an index between n_γ along X and n_β along the optic axis, and is the slower wave.

For optically positive crystals, all vibration components parallel to the optic plane are fast between isogyres on their convex side and slow between the field edge and an isogyre on its concave side. The reverse is, of course, true for optically negative crystals.

If a student will always orient the optic plane NE–SW, as shown in Plate IX, retardation will always subtract between the isogyres for positive crystals, and add between isogyres for negative ones.

OBTUSE BISECTRIX (Bxo) FIGURE (*Fig. 8-11*). An obtuse bisectrix interference figure is, of course, formed by a crystal section or fragment lying with its obtuse bisectrix vertical. It is a bisectrix figure with a very large axial angle (i.e., more than 90°), which places the melatopes far beyond the limits of view and causes isogyres to sweep swiftly through the field as the stage is rotated.

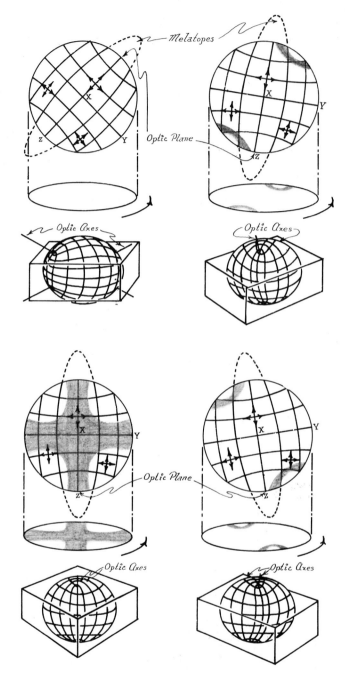

FIGURE 8-11. Formation of isogyres with stage rotation—obtuse bisectrix. Melatopes lie beyond the visible field, producing no visible isogyres at 45°, but a broad, diffuse cross in the 90° position. Note that isogyres enter and leave the field with relatively little stage rotation.

Plates

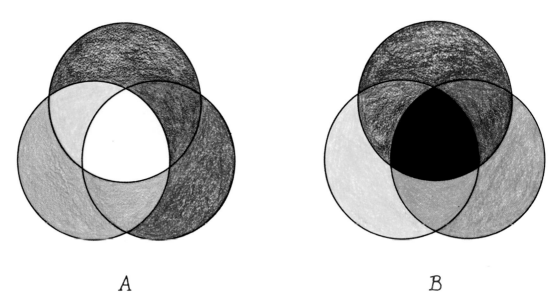

PLATE I. Light addition and light subtraction. (A) Three spotlights of primary light colors—red, green, and blue-violet—combine to produce white light by addition, and the complementary color of each primary light color is produced by addition where the other two primary light colors overlap. (B) The three circular areas of the primary pigment colors—red (magenta), yellow and blue (cyan)—printed on this page combine to produce black by absorbing all light. The complementary color of each primary pigment color is produced by combined absorption where the two other primary pigment colors overlap.

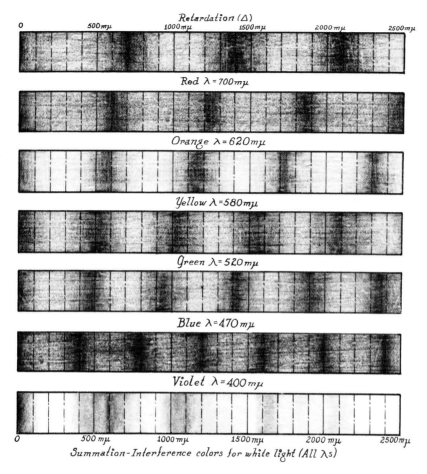

PLATE II. The summation of interference colors for all wavelengths of light. For any monochromatic light, darkness appears where increasing thickness causes retardation equal to some full multiple of the wavelength (i.e., $\Delta = n\lambda$). For white light (i.e., all wavelengths) the interference effect is a summation of the interference for all individual wavelengths. Where no wavelengths pass, the result is blackness; where all wavelengths pass, white light is seen; and where a large continuous segment of the spectrum is removed (e.g., violets and blues) the complementary color shows through (e.g., yellow plus red and green add to form yellow). With greater wedge thickness, the light and dark bands for different colors become staggered, with the result that several small scattered segments of the spectrum are removed and only scattered segments show through, adding to yield pale colors tending toward high-order gray or white.

PLATE III. Quartz wedge between crossed and parallel Polaroid sheets in monochromatic and white light. (A) In monochromatic (Na_D) light between crossed polars the quartz wedge shows brightness when $\Delta = (n + \frac{1}{2})\lambda$ and darkness when $\Delta = n\lambda$. (B) In monochromatic (Na_D) light between parallel polars the quartz wedge shows brightness when $\Delta = n\lambda$ and darkness when $\Delta = (n + \frac{1}{2})\lambda$. (C) In white light between crossed polars the quartz wedge shows a continuous sequence of colors resulting from the absence of those colors for which $\Delta = n\lambda$. (D) In white light between parallel polars the quartz wedge shows a continuous sequence of colors that are essentially complementary to those formed between crossed polars, and which result from the absence of those colors for which $\Delta = (n + \frac{1}{2})\lambda$.

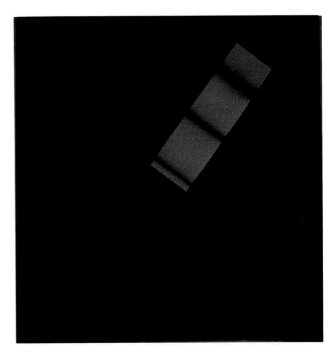

A

B

C

D

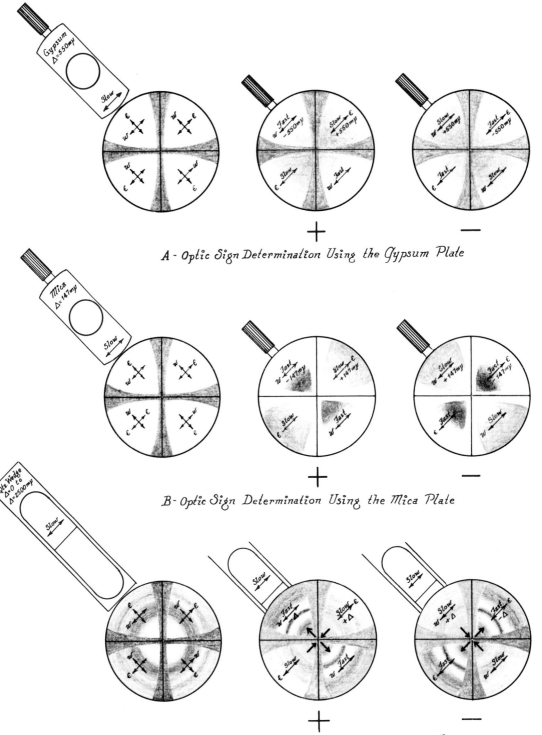

PLATE IV. Determination of optic sign from uniaxial optic-axis interference figures. A uniaxial crystal is positive if the ordinary rays are faster than the extraordinary rays, and negative if the extraordinary rays are faster. The use of an accessory plate or wedge superposes the slow vibration direction of the accessory on the vibration direction of the extraordinary waves (ϵ) in the upper-right and lower-left quadrants of the figure, where addition of retardation indicates a positive crystal. Subtraction of retardation in the upper-right and lower-left indicates a negative crystal, since extraordinary rays (ϵ) are faster. (A) The gypsum plate retards by 550 mμ those waves vibrating parallel to its slow vibration direction. It retards the ordinary waves in the NW and SE quadrants and the extraordinary waves in the NE and SW quadrants. If the original figure is largely first-order white ($\Delta \simeq 150$ mμ) and the crystal is positive, the gypsum plate causes the NW and SE quadrants to be yellow ($\Delta = 150 - 550$ mμ) because the ordinary wave is the faster wave, 150 mμ ahead of the extraordinary wave. In the NE and SW quadrants, where the slow extraordinary wave parallels the slow vibration direction of the gypsum plate, the retardation of the plate adds to that already present ($\Delta = 150 + 550$ mμ), and the field is blue. For negative crystals, the quadrant colors are reversed, because the extraordinary wave is faster. (B) The mica plate produces static retardation of 147 mμ. For positive crystals retardations add in the NE and SW quadrants and subtract in the NW and SE. Colors are reversed for negative crystals. (C) As the quartz wedge is inserted, thin end first, colors flow evenly in the directions of the short, heavy arrows. The quartz wedge is best used when an interference figure shows many isochromes.

PLATE V. Determining the sign of an off-center uniaxial optic-axis figure. With the melatope beyond the visible field, only one isogyre or one quadrant is seen at a given position of stage rotation. One must visualize the melatope position and determine the optic sign as though the entire optic-axis figure were visible.

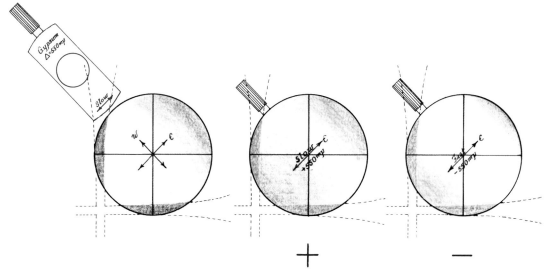

Uniaxial, Optic Axis, Off Center Figure - Sign Determination

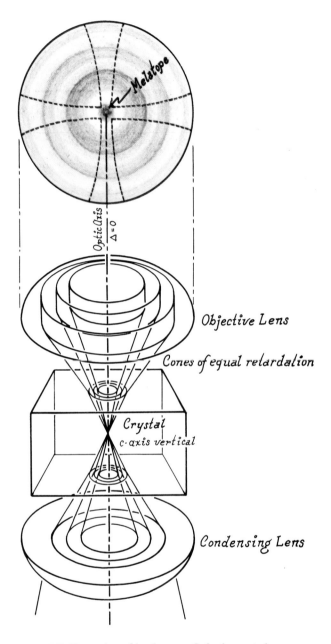

PLATE VI. Formation of isochromes. Only the central ray passes along the optic axis with zero retardation. Since all other rays are inclined to the optic axis, retardation increases with inclination from the central ray. Note that both thickness (d) and birefringence ($n_2 - n_1$), and hence retardation, increase uniformly from the center outward. Diverging cones of equal retardation yield rings of equal interference color centered on the optic axis.

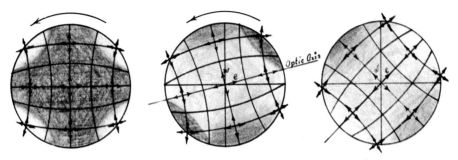

A — Uniaxial Flash Figure — Locating the Optic Axis

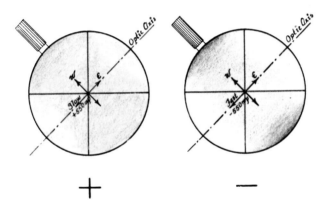

B — Uniaxial Flash Figure — Sign Determination with Gypsum Plate

PLATE VII. Determining the sign of a uniaxial flash figure. A flash figure forms when the optic axis lies in the plane of the microscope stage. (A) When the optic axis is N–S or E–W, the field is darkened by a broad, diffuse isogyre that moves rapidly toward the optic axis and leaves the visible field within only a few degrees of stage rotation. (B) At 45° the optic axis lies in the quadrants of low birefringence, and an accessory plate can be used to reveal whether the extraordinary wave is slow or fast.

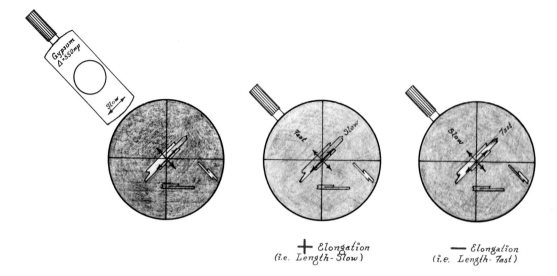

PLATE VIII. Determining the sign of elongation. Uniaxial crystals or cleavage fragments that show consistent elongation must be elongated parallel to *c*. An accessory plate may be used to determine whether the slow or fast wave vibrates parallel to the direction of elongation.

PLATE IX. Determination of optic sign from biaxial acute bisectrix figures. A biaxial crystal is positive if Z is the acute bisectrix and negative if X bisects the $2V$ angle. Accessory plates are used to determine whether the central wave vibrating in the optic plane is fast or slow compared to the wave vibrating parallel to the optic normal Y. (A) The gypsum plate causes the wave parallel to the optic plane to be slowed by 550 mμ, causing retardation between the isogyres to subtract for positive crystals and to add for negative ones. (B) The mica plate causes the wave parallel to the optic plane to be retarded by 147 mμ. (C) Insertion of the quartz wedge causes color bands to move in the direction indicated by the heavy arrows.

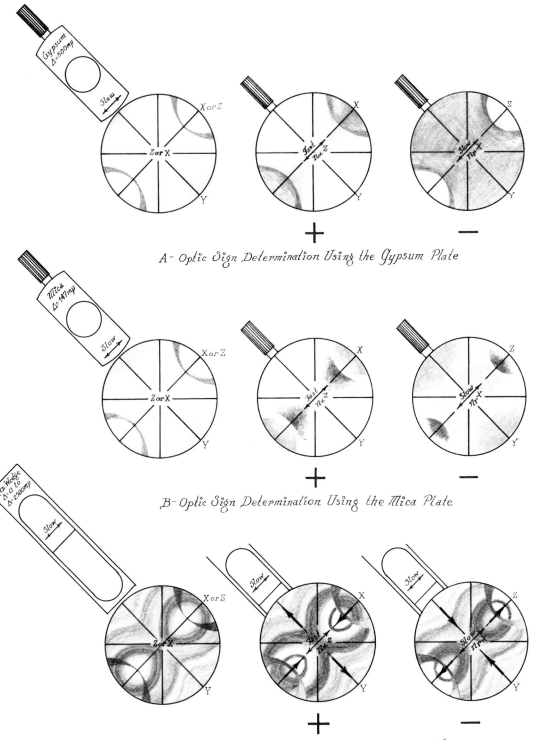

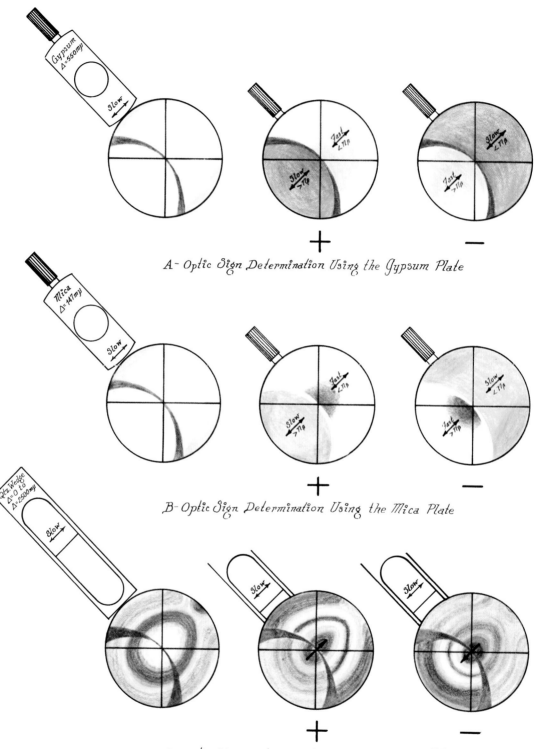

A – Optic Sign Determination Using the Gypsum Plate

B – Optic Sign Determination Using the Mica Plate

C – Optic Sign Determination Using the Quartz Wedge

PLATE X. Determination of optic sign from biaxial optic-axis figures. An optic-axis figure may be considered as one-half of an acute bisectrix. Accessory plates serve to indicate whether the wave vibrating in the optic plane is fast or slow compared to the wave vibrating parallel to the optic normal. (A) The gypsum plate retards or advances existing interference colors by 550 mμ. (B) The mica plate retards or advances existing interference colors by 147 mμ. (C) Insertion of the quartz wedge causes color bands to move in the direction shown by heavy arrows. Remember that the acute bisectrix lies on the convex side of the isogyre and that if the isogyre seems to have no curvature in its 45° position the 2V angle is near 90° and the crystal is optically neutral.

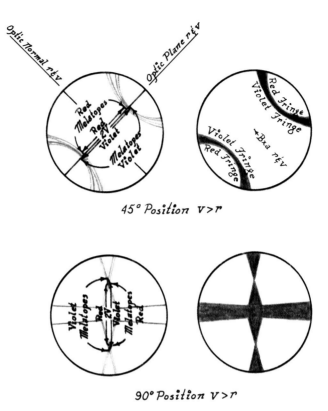

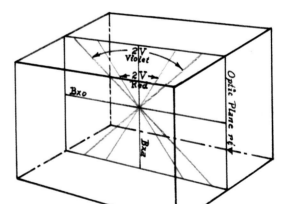

PLATE XI. Optic-axis dispersion. Optic-axis dispersion may be produced by orthorhombic, monoclinic, or triclinic crystals, and occurs whenever the 2V angle for long wavelengths (red) differs from that for short wavelengths (violet). Orthorhombic crystals produce only optic-axis dispersion. Monoclinic and triclinic crystals produce optic-axis dispersion alone or in combination with some type of bisectrix dispersion.

PLATE XII. Inclined bisectrix dispersion. Monoclinic crystals may produce inclined bisectrix dispersion when $Y = b$. Color fringes are symmetrical about the optic plane.

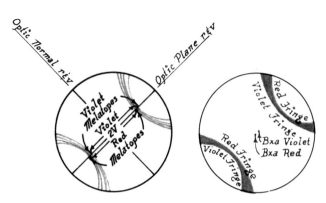

45° Position $r = v$

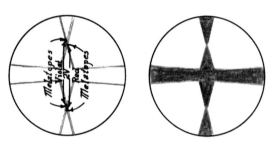

90° Position $r = v$

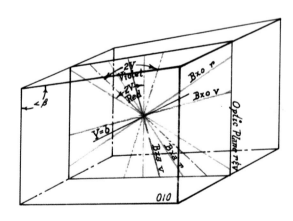

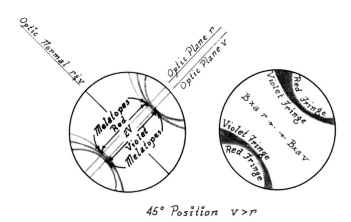

45° Position v > r

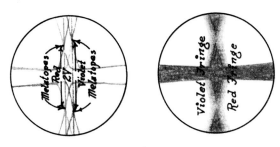

90° Position v > r

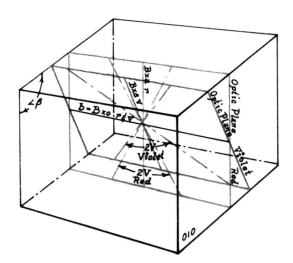

PLATE XIII. Horizontal bisectrix dispersion. Monoclinic crystals may produce recognizable horizontal dispersion when the obtuse bisectrix parallels *b*. Color fringes are symmetrical about the optic normal.

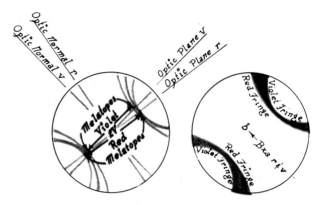

45° Position r>v

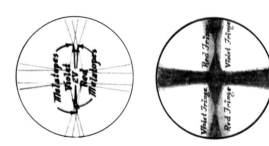

90° Position r>v

PLATE XIV. Crossed bisectrix dispersion. Monoclinic crystals may produce recognizable crossed dispersion when the acute bisectrix parallels *b*. Color fringes are symmetrical about the acute bisectrix.

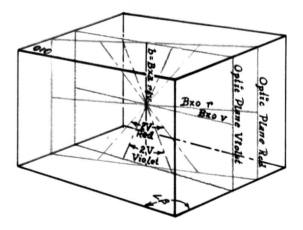

The distance between melatopes is again a measure of the axial angle, which must now exceed 90°; although melatopes are not visible in an obtuse bisectrix figure, the method of Michel-Lévy remains valid. The bisectrix figure is almost certainly acute if the isogyres never leave the outer limits of the visible field, used as a reference circle, or if more than 30° of stage rotation is required to separate the isogyres from their position of contact at the center of the field and move them to the edge of the visible field. If less than 15° of stage rotation separates the isogyres from their position of contact and moves them beyond the field limits, the bisectrix is obtuse*; if the isogyres disappear within 15° to 30° of stage rotation, the $2V$ is large (near 90°), and the visible bisectrix may be either acute or obtuse.

The student should not be discouraged when he encounters a crystal whose optic sign cannot be determined. For axial angles near 90° this is difficult; instead of being discouraged, he should appreciate the advantage he has gained by knowing that $2V$ is near 90°.

The problem of sign determination again consists in determining the relative velocities of the two mutually perpendicular waves advancing along the bisectrix, a procedure identical to that for an acute figure. Retardation affects are reversed, however, since the opposite bisectrix is involved. If a student will always orient the optic plane NE–SW, retardation will always add between isogyres for positive crystals and subtract for negative ones. Note that when the optic plane is 45° to the cross hairs, the entire visible field lies between isogyres. The student is cautioned to rotate the stage 45° away from isogyre contact (i.e., to maximum separation) before attempting a sign determination.

OPTIC AXIS FIGURE (*Fig. 8-12*). When the central ray passes along an optic axis, it experiences zero birefringence and forms a melatope at the center point of an isogyre. For $2V$ angles less than about 30°, a second melatope is visible with its accompanying isogyre, and for very small axial angles, the distinction between acute bisectrix and optic axis figures loses significance.

For larger axial angles, however, a single isogyre remains stationary at the melatope, with its two arms alternately making short angular sweeps opposite to the direction of stage rotation (Fig. 8-12). A line that passes through the melatope and symmetrically bisects the isogyre is the trace of the optic plane. When the optic plane is 45° to the cross hairs, the isogyre shows maximum parabolic curvature. The acute bisectrix lies on the trace of the optic plane, within or beyond the visible field, on the convex side of the parabolic isogyre.

---

*Flash figures, both uniaxial and biaxial, are essentially bisectrix figures with axial angles of 180° and are thus often confused with large-angle obtuse bisectrix figures.

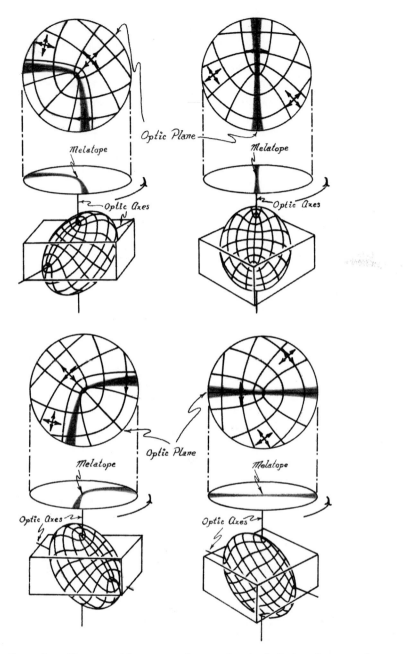

FIGURE 8-12. Formation of isogyres with stage rotation—optic axis. A single melatope at the center of the field remains stationary with stage rotation. The single isogyre becomes straight when the optic plane is N–S or E–W. At 45° positions, alternate halves of the bent isogyre make short sweeps in a direction opposite to stage rotation. The amount of curvature at 45° is inversely related to $2V$, and for $2V <$ about 30°, a second melatope and isogyre will be present in the visible field to form an off-center Bxa.

The degree of curvature is determined by the size of the axial angle (Fig. 8-8); if the isogyre shows inconclusive curvature, $2V$ is near $90°$, and any sign determination is inconclusive.

For sign determination, an optic-axis figure may be treated as half an acute bisectrix figure. If a student will always orient the optic plane NE–SW, as shown in Plate X, retardation will always add on the concave side of the isogyre for positive crystals and subtract on the concave side for negative ones.

FLASH FIGURES (*Fig. 8-13*). Biaxial flash figures result when the optic normal (Y) is vertical. They cannot be distinguished from uniaxial flash figures (Plate VII) and are often confused with obtuse bisectrix figures. When X and Z parallel the cross hairs, we see a very broad, diffuse, dark cross that separates quickly into two diffuse, hyperbolic isogyres that leave the field in the direction of the acute bisectrix with slight stage rotation. If $2V$ is small and birefringence not too low, the field will be bright at $45°$, and alternate quadrants will show lower retardation and thus mark the position of the acute bisectrix. For large $2V$, retardation in all four quadrants is equal, and no distinction can be made between the acute and the obtuse bisectrix. If the position of the acute bisectrix can be oriented NE–SW, any accessory plate will aid in determining whether the slower wave (i.e., Z) parallels the acute or the obtuse bisectrix.

OFF-CENTER FIGURES (*Fig. 8-14*). When neither X, Y, Z nor an optic axis is vertical, the resulting interference figure is an off-center equivalent of one or more of the types previously discussed. Unless one or both melatopes appear in the visible field, an interference figure can yield only limited useful information; consequently, it is usually worth the effort to find a better figure. Nevertheless, with the assumption that even the worst figure has something to offer a keen observer, the following three short paragraphs are offered.

When Y lies in the plane of the microscope stage, the optic plane must symmetrically divide the field of view, and if neither X, Z, nor an optic axis is vertical, the melatope(s) of the resulting off-center optic-axis figure will occupy some position other than the central point (Fig. 8-14,A). When a bisectrix lies in the plane of the stage, the trace of the optic normal must symmetrically divide the field (Fig. 8-14,B).

Interference figures of the type shown in Figure 8-14,A and B are suggestive of $\{hk0\}$, $\{h0l\}$, or $\{0kl\}$ sections of orthorhombic crystals or $\{h00\}$, $\{00l\}$, or $\{h0l\}$ sections of monoclinic crystals.

When neither X, Y, nor Z lies in the plane of the stage, we see a truly oblique figure in which neither the trace of the optic plane nor optic normal symmetrically bisects the field of view (Fig. 8-14,C). Such figures suggest $\{hkl\}$

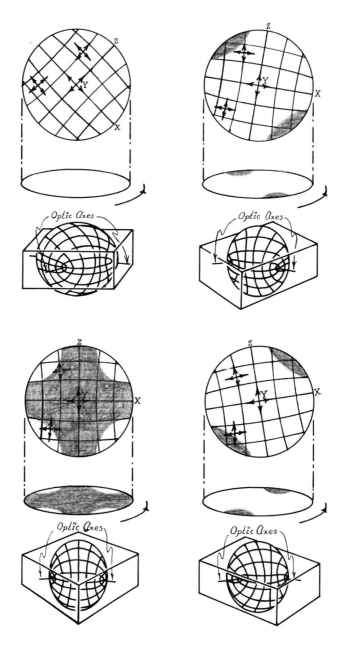

FIGURE 8-13. Formation of isogyres with stage rotation—flash figure. For crystals of large $2V$, vibration patterns of waves traveling near the optic normal form an almost square grid, producing momentary total extinction every 90° of stage rotation. When $2V$ angles are less than 90°, curvature in the vibration pattern produces a very broad, diffuse cross at 90° which enters and leaves the field with only a few degrees of stage rotation. As $2V$ becomes small, the flash figure closely resembles the Bxo, and at $2V = 0°$ the distinction is lost between optic normal and Bxo (uniaxial flash figure).

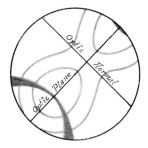 

A - Optic Normal (Y) in plane of microscope stage

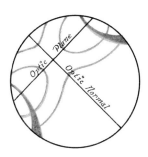 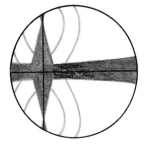

B - Obtuse Bisectrix in plane of microscope stage

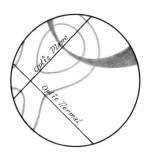 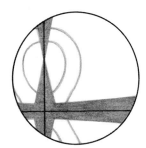

C - Neither X, Y nor Z in plane of microscope stage

FIGURE 8-14. Off-center biaxial figures. Off-center figures result when neither X, Y, Z nor an optic axis is centered in the figure (i.e., normal to the visible field). (A) When Y lies in the plane of the microscope stage, the resulting interference figure is symmetrical about the optic (XZ) plane. (B) When a bisectrix lies in the plane of the stage, the figure is symmetrical about the optic normal. (C) When neither X, Y, nor Z lie in the plane of the stage, the orientation is truly random, and the figure is symmetrical about neither optic normal nor optic plane.

sections of orthorhombic crystals, {hkl}, {hk0}, and {0kl} sections of monoclinic crystals, or almost any section of a triclinic crystal.

Dispersion Effects in Interference Figures

Dispersion results whenever long waves and short waves follow separate paths. It is usually described by such questionable terms as imperceptible, slight, weak, distinct, moderate, strong, and extreme. Although often "imperceptible," or easily overlooked, dispersion is an optical property of all transparent media, isotropic or anisotropic, uniaxial or biaxial; it is also independent of other optical properties and is difficult or impossible to predict from composition or structure.*

For an isotropic medium, dispersion implies that rays at one end of the visible spectrum advance faster than those at the other, and thus produce spherical ray velocity surfaces of different radii.

For uniaxial crystals, ray velocity, and hence n_ω and n_ϵ, differ for red light from those for violet light. The indicatrix for red light may differ in size and even in shape from that for violet light, but since crystallographic orientation never changes, dispersion is not observable in uniaxial interference figures.

OPTIC-AXIS DISPERSION (*Fig. 8-15,A*). In biaxial minerals, dispersion may cause refractive index to differ with wavelength, and since this change may be greater for one principal index than for another, the shape of the indicatrix and the angular separation of the optic axes may vary continuously with wavelength. *Dispersion that causes the magnitude of the 2V angle to vary with wavelength is called optic-axis dispersion.* It is described by the notation "$r > v$" when the 2V for red light is greater than 2V for violet light and by "$v > r$" for the converse relation.

BISECTRIX DISPERSION (*Fig. 8-15,B*). Where orientation of the optical directions X, Y, and Z is not rigidly established by crystal symmetry (i.e., in monoclinic and triclinic crystals), dispersion may cause the angular relationship between optical and crystallographic directions to vary with wavelength. *Dispersion that causes optical directions to vary with wavelength is called bisectrix dispersion.*

Bisectrix dispersion is necessarily absent in orthorhombic crystals, since optical directions must parallel cystallographic directions irrespective of wave-

---

*Many phosphate and sulfate minerals and many minerals containing heavy elements show strong dispersion.

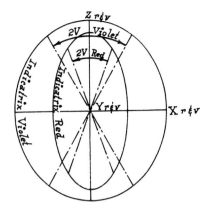
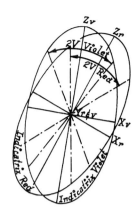

A - *Optic Axis Dispersion (v > r)*   B - *Bisectrix Dispersion*

FIGURE 8-15. Cause of basic types of dispersion. (A) Optic-axis dispersion results when the size of the $2V$ angle varies with wavelength. The values of the principal indices n_α, n_β, and n_γ, the shape of the indicatrix, and the $2V$ angle all vary with wavelength. (B) Bisectrix dispersion results when the orientation of the indicatrix with respect to crystallographic directions varies with wavelength. The quantitative measure of either type may range from "extreme" to "imperceptible," and each may show several qualitative types.

length.* In monoclinic crystals, bisectrix dispersion takes several forms, depending on which optical direction is fixed parallel to the *b*-crystallographic axis. Triclinic symmetry allows additional forms of bisectrix dispersion, since no optical directions are fixed. Under favorable conditions, *bisectrix dispersion thus offers a method of establishing crystal system.*

DISPERSION IN ORTHORHOMBIC INTERFERENCE FIGURES (*Plate XI*). *Orthorhombic symmetry permits only optic-axis dispersion*, and the $2V$ angle for red light may be greater or less than that for violet. Herein is a useful, independent, optical property that is observable in acute bisectrix figures,† where the melatopes

---

*A very rare but interesting exception is the orthorhombic mineral brookite. For red light brookite shows its maximum refractive index (n_γ) for light waves vibrating parallel to the *b*-crystallographic axis ($b = Z$), its intermediate principal index (n_β) for waves parallel to $c(c = Y)$, and its minimum index (n_α) for waves parallel to $a(a = X)$. For violet light, $b = Z$, $c = X$, and $a = Y$. The optic plane is the *ab*-plane for red light and the *cb*-plane for violet; consequently, brookite is said to show crossed axial-plane dispersion.

†Optic-axis dispersion is also readily apparent in optic-axis figures and, at least theoretically, in obtuse bisectrix figures. An optic-axis figure may, however, not reveal the presence or absence of accompanying bisectrix dispersion.

and isogyres for one end of the visible spectrum are more widely spaced (i.e., larger $2V$) than those for the other end. The melatopes (and isogyres) for red light are sites where red light is absent and opposite wavelengths (i.e., violet) are visible. Where isogyres for all wavelengths overlap, all wavelengths are absent and the isogyre is black. When dispersion spreads the spectral colors, we see a delicate fringe of short wavelengths (i.e., violet) on the red melatope side of the black isogyre and a fringe of long wavelengths (i.e., red) on the violet melatope side. *When violet fringes are the more widely spaced (i.e., on the concave side of isogyres), $2V$ for red is greater than $2V$ for violet ($r > v$); conversely, red fringes on the concave side indicate that $v > r$* (Plate XI).

Orthorhombic dispersion produces spectral color fringes that range from extreme to imperceptible and which are symmetrical about both the trace of the optic plane and the optic normal in both 90° and 45° positions. Color fringes largely disappear when the optic plane is N–S or E–W, because melatopes for all colors lie on an isogyre and the isogyres for all colors are superimposed.

DISPERSION IN MONOCLINIC INTERFERENCE FIGURES. Monoclinic symmetry permits both optic-axis and bisectrix dispersion, and requires that one of the three optical directions always be parallel to b. Thus the following three types of bisectrix dispersion are possible.

1. *Inclined bisectrix dispersion results when the optic normal (Y) parallels the b-crystallographic axis* (Plate XII). The optic plane (XZ) must parallel the single crystal symmetry plane, but the position of X, Z, and the optic axes within this plane change with wavelength. If optic-axis dispersion is slight (i.e., $r \simeq v$), inclined dispersion of the bisectrices spreads the spectral colors to produce alternating red and violet color fringes (see Plate XII). Strong, superimposed optic-axis dispersion may displace the color fringes so that fringes of similar color appear on similar sides of both isogyres, but inclined bisectrix dispersion should cause color fringes to be stronger on one isogyre. *Inclined bisectrix dispersion produces spectrial color fringes symmetrical about the trace of the optic plane but not about the optic normal.* Color fringes essentially disappear in the 90° position.

2. *Horizontal, or parallel, bisectrix dispersion seen in the acute bisectrix figure, results when the obtuse bisectrix (X or Z) parallels the b-crystallographic direction* (Plate XIII). Change in wavelength may cause rotation of the optic plane about the obtuse bisectrix. An acute bisectrix figure shows parallel displacement of the optic planes for red and violet light (see Plate XIII). Superimposed optic-axis dispersion may cause separation of red and violet isogyres

along the direction of the optic planes ($r > v$ or $v > r$); horizontal bisectrix dispersion causes the red and violet isogyres to be displaced along the direction of the optic normal.

Horizontal bisectrix dispersion produces spectral color fringes that are apparent in both 45° and 90° positions and which are symmetrical about the optic normal but not about the trace of the optic plane. Optic-axis dispersion is best observed when the isogyres show maximum separation (i.e., 45° position); horizontal bisectrix dispersion is most easily recognized in the 90° position, since similar color fringes occur in adjacent quadrants on the same side of the optic plane.

3. *Crossed, or rotated, bisectrix dispersion, seen in the acute bisectrix figure, results when the acute bisectrix (Z or X) parallels the b-crystallographic direction* (Plate XIV). Change in wavelength may cause rotation of the optic plane about the fixed acute bisectrix. An acute bisectrix figure* shows crossed optic planes for red and violet light intersecting at the bisectrix (Plate XIV). Superimposed optic-axis dispersion may cause separation of red and violet isogyres along the optic plane ($r > v$ or $v > r$); bisectrix dispersion spreads the isogyres normal to the optic plane in a crossed relationship.

Crossed bisectrix dispersion produces color fringes that are symmetrical about the acute bisectrix and reproduced by two-fold (180°) rotation, but symmetrical about neither the optic plane nor the optic normal. Color fringes are present in both 90° and 45° positions; crossed bisectrix dispersion is most apparent when the optic plane is N–S or E–W. Similar color fringes appear in alternate quadrants.

DISPERSION IN TRICLINIC INTERFERENCE FIGURES. Triclinic symmetry permits both optic-axis and bisectrix dispersion with no restrictions upon orientation. Changes in wavelength may cause changes in the crystallographic orientation of all three optical directions, and triclinic bisectrix dispersion may produce spectral *color fringes that show no symmetry about optic plane, optic normal, or bisectrix* and which appear in both 90° and 45° positions.

Unfortunately, theoretical dispersion effects are much easier to draw than they are to observe; students should therefore be alerted that when dispersion is weak the *theoretical* symmetry for bisectrix dispersion might possibly be lower than what is observed (i.e., triclinic dispersion may appear monoclinic, and monoclinic dispersion may appear orthorhombic).

---

*Note that when an acute bisectrix figure shows crossed dispersion, the obtuse bisectrix shows horizontal dispersion and vice versa. Similarly, if an acute bisectrix shows optic-axis dispersion with $r > v$, the obtuse bisectrix shows optic-axis dispersion with $v > r$.

The Search for Interference Figures

Crystal fragments or sections that show minimum birefringence yield optic-axis figures that are most useful for determination of optic sign and rapid estimation of $2V$, but dispersion is best seen in acute bisectrix figures.

The isogyres of an off-center biaxial optic-axis figure tend to sweep across the visible field 45° to the cross hairs, whereas those of off-center uniaxial optic-axis figures tend to sweep through the field parallel to the cross hairs; unless one melatope lies within the visible field, however, meaningful observations are limited.

Sections or fragments of monoclinic and triclinic crystals with high dispersion show dispersion colors superimposed on retardation colors. The effect is especially apparent on grains of minimum birefringence, which have no positions of complete extinction and appear alternately red and blue with stage rotation because extinction positions differ for red and violet wavelengths.

Index of Refraction

The Search for Principal Indices

Principal indices are determined from immersed fragments by (1) selection between crossed nicols of properly oriented fragments and (2) index measurement between uncrossed nicols by comparison with immersion liquids. The latter process was covered in Chapter 3; this section is confined to its practical application to biaxial crystals.

Fragments showing minimum birefringence have a vertical optic axis, a circular indicatrix section, and *an index of n_β*, regardless of the stage rotation. *Fragments showing maximum birefringence ($n_\gamma - n_\alpha$) will yield n_α in one extinction position and n_γ in the other.* Other principal sections show indices of n_α and n_β or n_β and n_γ. Crystals in such orientations, however, are difficult to recognize, except where acute bisectrix and obtuse bisectrix interference figures make them known. Randomly oriented sections must show one index between n_α and n_β in one extinction position, and another index between n_β and n_γ in the other.

Fragments of minimum birefringence are easily recognized with a low-power objective lens. Fragments of maximum birefringence, however, are more difficult to recognize; the student will therefore do well to select several highly birefringent fragments for index measurements, with the foreknowledge that no fragment can show an index above n_γ or below n_α. If cleavage is such that fragments do not lie in useful orientations, the gelatin-coated slides described on page 62 may be used, or mineral fragments may be mixed with glass fragments to encourage random orientation.

Color and Pleochroism

Like uniaxial and isotropic crystals, biaxial crystals are colored by selective light absorption, and transition elements are largely responsible for mineral coloration. Waves vibrating parallel to optical directions X, Y, and Z experience three principal refraction indices, n_α, n_β, and n_γ, and three independent types and degrees of light absorption. Three-fold pleochroism is termed *trichroism*, and is expressed by a formula analogous to that for uniaxial pleochroism; for example,

$$n_\alpha = \text{dark red brown,}$$
$$n_\beta = \text{pale red brown,}$$
$$n_\gamma = \text{pale yellow brown.}$$

Or "quantitative" symbolism may read:

$$X > Z > Y,$$

indicating that waves vibrating parallel to X are absorbed most strongly and those parallel to Y are absorbed least. No attempt is made to quantify pleochroism beyond use of the usual descriptive terms that express the range from imperceptible to extreme.

Crystal Orientation

Relationship Between Optical and Crystallographic Directions

Symmetry planes of the indicatrix (i.e., principal sections) must parallel crystal symmetry planes in so far as the latter exist for the normal class of orthorhombic, monoclinic, and triclinic crystals.

For orthorhombic crystals, optical directions must parallel crystallographic directions, and the biaxial indicatrix has six possible orientations:

$$X = a, \quad Y = b, \quad Z = c,$$
$$X = a, \quad Y = c, \quad Z = b,$$
$$X = b, \quad Y = a, \quad Z = c,$$
$$X = b, \quad Y = c, \quad Z = a,$$
$$X = c, \quad Y = a, \quad Z = b,$$
$$X = c, \quad Y = b, \quad Z = a.$$

For monoclinic crystals, one optical direction must parallel the *b*-crystallographic direction, and the remaining two optical directions lie in the *ac*-crystallographic plane (i.e., {010}). The three possible orientations are:

$$X = b,$$
$$Y = b,$$
$$Z = b.$$

The two remaining optical directions may lie anywhere within the *ac*-plane. To describe the optical orientation of a given monoclinic crystal, we must note which optical direction parallels *b* and designate an angle between one optical and one crystallographic direction within the *ac*-plane. This angle is the *orientation angle*, and is usually written $c \wedge Z$, $c \wedge Y$, $a \wedge X$, etc., and read "the angle from *c* to *Z*" or simply "*c* angle *Z*." If the *ac*-plane (i.e., the 010 section) is represented with the *a*-axis dipping downward to the left across a vertical *c* (Fig. 8-16), the orientation angle is positive when measured clockwise *from the crystallographic direction*, usually *c*, to the optical direction, and is negative when measured counterclockwise.

Most monoclinic minerals are characterized by greater or lesser amounts of ionic substitution, which is in turn responsible for variation in the orientation angle, causing it to be expressed as an allowable range rather than a simple value. This angle is also a function of bisectrix dispersion and can be a constant value only for a given wavelength and a given composition.

For triclinic crystals, no predictable relationship exists between optical and crystallographic directions; consequently, three orientation angles are necessary to fix the indicatrix position. Triclinic minerals are comparatively rare, and most common ones are essentially monoclinic, as one principal indicatrix section lies within a few degrees of a pinacoidal plane.

Extinction Angle

The angle between an extinction position of a grain or fragment and some crystallographic direction, as indicated by cleavage traces, crystal faces, or twinning planes, is its extinction angle. For uniaxial crystals, extinction is

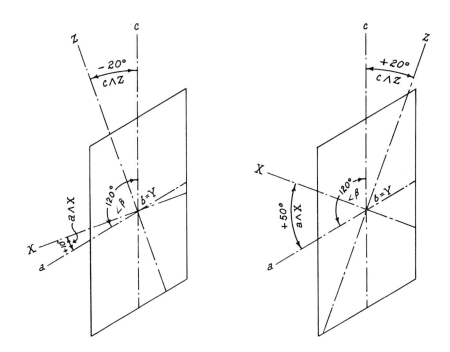

FIGURE 8-16. Orientation angle. When the *ac*-plane of a monoclinic crystal is represented with *a* dipping left, the orientation angles are positive when measured clockwise from the crystallographic direction and negative when measured counterclockwise.

either parallel or symmetrical. Some biaxial crystal planes also show parallel or symmetrical extinction, but the characteristic extinction for low-symmetry crystal systems is inclined (Fig. 8-17,C); between crossed nicols, the crystal extinguishes when cleavage or other visible crystal planes lie at an angle to the cross hairs.

Orthorhombic crystals show only *parallel or symmetrical extinction*. Extinction must occur parallel to all pinacoidal planes {100}, {010}, or {001}; symmetrical to pyramidal planes {hkl}; and either parallel or symmetrical to prism {hk0} or dome {h0l} or {0kl} planes.

Monoclinic crystal fragments or sections may show *parallel, symmetrical, or inclined extinction*, depending on the orientation. When our line of observation is normal to the *b*-crystallographic axis (i.e., within the {010} plane) extinction is parallel to pinacoidal planes and parallel or symmetrical to prismatic planes. In all directions of observation outside the {010} plane, extinction is inclined to all crystal planes; the maximum angle is observed normal to {010}. Maximum extinction angle is a measure of the orientation angle (e.g., $c \wedge Z$) and

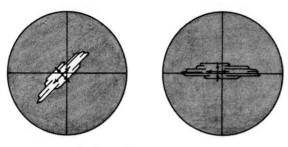

A - Parallel Extinction

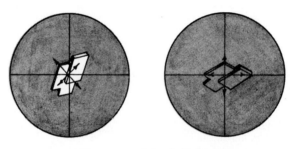

B - Symmetrical Extinction

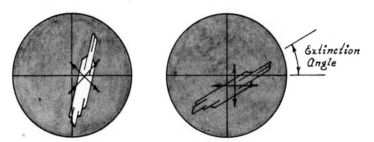

C - Inclined Extinction

FIGURE 8-17. Extinction of biaxial crystal cleavages. Sections or fragments of biaxial crystals may possess cleavages which are parallel (A), symmetrical (B), or inclined (C) to the vibration directions of transmitted light waves.

has a characteristic value, or range of values, for each monoclinic mineral. Since it may be difficult to know when observation is truly normal to {010} it may be necessary to measure extinction angles on a number of crystal fragments or sections to establish a maximum value.

Triclinic crystal fragments and grains show *only inclined extinction*, as a consequence of optical-crystallographic orientations. In practice, it is possible to observe nearly parallel or even symmetrical extinction through a combination of accidental coincidence of crystallographic and optical directions and the essentially monoclinic nature of many triclinic minerals.

Sign of Elongation

Crystal habit and cleavages may produce elongated fragments and grains that show lineation related to polarization planes of fast and slow waves by extinction angle and sign of elongation. Biaxial elongation with parallel or near-parallel extinction may be described in a manner analogous to that used for uniaxial lineation (p. 118), but inclined extinction requires re-evaluation of the elongation sign.

Fragments and sections of orthorhombic crystals normally show an optical direction X, Y, or Z parallel to elongation, producing parallel extinction. When X lies along lineation, the orientation is always length-fast (i.e., elongation is negative), since the fast wave vibrates parallel to lineation. Similarly, elongation parallel to Z can only be length-slow (i.e., elongation is positive). Elongation parallel to Y, however, may be either positive or negative, depending on orientation about Y.

Fragments and sections of monoclinic and triclinic crystals normally show inclined extinction where polarization planes are not parallel to elongation. If the extinction angle is small, elongation is positive or negative; but when extinction approaches 45°, elongation lies halfway between fast and slow waves, and the sign elongation becomes insignificant. For extinction angles less than 45°, the vibration plane of either the fast or the slow wave is nearer elongation.

Cleavage of Biaxial Crystals

Description of all possible cleavages of biaxial crystals and their relationship to optical directions is a much greater undertaking than the analogous description of simple basal, prismatic, and pyramidal cleavages of uniaxial crystals. The number of possible, unrelated cleavage directions increases rapidly as crystal symmetry decreases, and the influence of cleavage on the optical study of biaxial crystals is perhaps best illustrated by a few specific examples.

Orthorhombic minerals most commonly display cleavage parallel to pinacoids or prisms. Each pinacoidal cleavage {001}, {100}, or {010} is independent, but prismatic cleavage {hk0} requires two cleavage directions.

Topaz (Fig. 8-18) is an optically positive, orthorhombic mineral with a single perfect pinacoidal cleavage {001} and orientation $Z = c$, $X = a$, and $Y = b$. Fragments of topaz may be expected to lie on {001} cleavage planes; to show no consistent elongation, and hence no measureable extinction angle or sign of elongation; and to yield acute bisectrix interference figures with only optic-axis dispersion and refraction indices n_α and n_β in extinction positions. Sections of topaz in which the single cleavage is vertical must show parallel extinction and negative elongation with respect to the cleavage and must yield centered or off-center flash figures or obtuse bisectrix figures and a refraction index of n_γ when cleavage is E–W and an index of n_β to n_α when cleavage is N–S.

Triphylite (Fig. 8-19) is an optically negative orthorhombic mineral with orientation $Z = b$, $Y = a$, and $X = c$ and cleavages parallel to {001} (perfect) and {010} (distinct). Cleavage fragments of triphylite should be elongated parallel to the a-axis, and should lie on either {001} or {010}. Those lying on {001} show parallel extinction and negative elongation and yield an acute

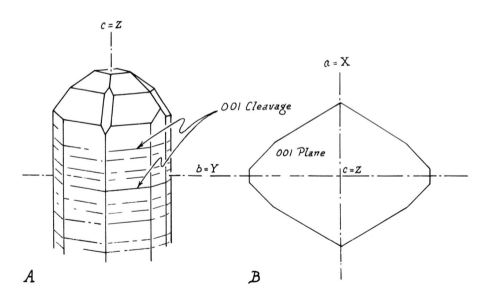

FIGURE 8-18. Optical orientation of topaz. (A) Topaz is orthorhombic with $a = X$, $b = Y$, and $c = Z$ and perfect cleavage parallel to {001}. (B) Single cleavage fragments lie on {001}.

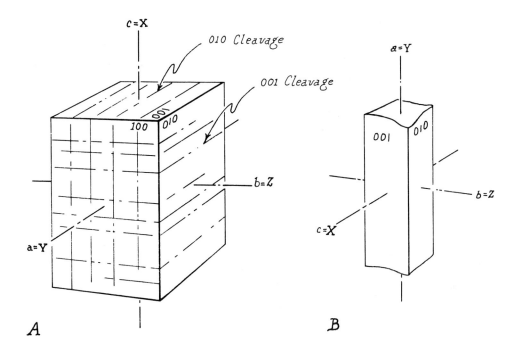

FIGURE 8-19. Optical orientation of triphylite. (A) Triphylite is orthorhombic with $a = Y$, $b = Z$, and $c = X$ and cleavage parallel to $\{001\}$ and $\{010\}$. (B) Single cleavage fragments are elongated on $a = Y$ and may lie on either $\{001\}$ or $\{010\}$.

bisectrix figure having only optic-axis dispersion. They give a refractive index of n_β when elongated N–S and n_γ when elongated E–W; birefringence is $(n_\gamma - n_\beta)$. Those fragments lying on $\{010\}$ show parallel extinction and positive elongation and yield an obtuse bisectrix figure. They give refractive index n_β when oriented N–S and n_α when oriented E–W; birefringence is $(n_\beta - n_\alpha)$. Crystal sections of triphylite also have predictable properties dependent upon orientation. Sections parallel to $\{001\}$ and $\{010\}$ are described above, and principal section $\{100\}$ shows two cleavages at right angles parallel to extinction directions and yields flash figures and maximum birefringence.

Monoclinic minerals commonly show cleavage parallel to prisms, as in amphiboles and pyroxenes; parallel to the basal pinacoid, as in micas, chlorites, and other layered silicates; and parallel to other pinacoids, as in orthoclase and gypsum. Cleavages are independent except prismatic $\{hk0\}$, which requires two cleavage directions.

Muscovite (Fig. 8-20) is monoclinic (β angle $\simeq 92°$) and optically negative with perfect $\{001\}$ cleavage, orientation $b = Z$ and $c \wedge X = 1°$ to $3°$. Cleavage fragments lie on $\{001\}$ with the acute bisectrix X about $3°$ from the vertical. Fragments, or sheets, show no cleavage for extinction angle or orientation and yield slightly off-center acute bisectrix figures with horizontal bisectrix dispersion. Birefringence is essentially $(n_\gamma - n_\beta)$, and the fragment shows a high refraction index n_γ and a lower index slightly less than n_β in extinction positions.

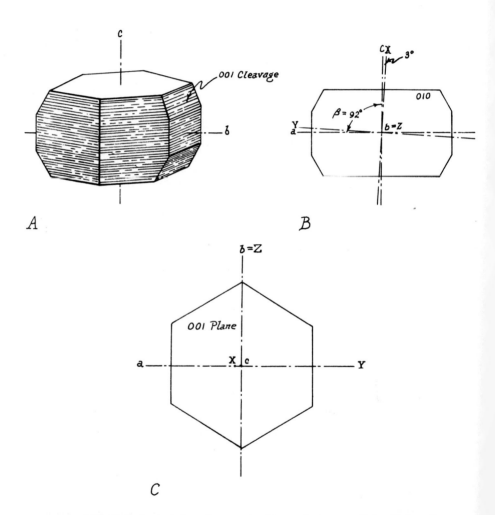

FIGURE 8-20. Optical orientation of muscovite. Muscovite is monoclinic with $b = Z$, $c \wedge X = +3°$, $\angle \beta = 92°$, and perfect basal $\{001\}$ cleavage. Cleavage sheets (C) lie with X about $3°$ from vertical.

Sections normal to the cleavage give an extinction angle ranging from 0° for sections parallel to $b = Z$ to 1°–3° for {010} sections. They always show positive elongation, range in birefringence from $n_\beta - n_\alpha$ for {010} sections to essentially $n_\gamma - n_\alpha$ for others, and yield almost worthless interference figures between the obtuse bisectrix and the optic normal.

Hornblende (Fig. 8-21) offers a classical example of a monoclinic mineral with perfect prismatic {110} cleavages. Assume a specific example where $n_\alpha = 1.634$, $n_\beta = 1.676$, $n_\gamma = 1.688$, orientation $b = Y$ and $c \wedge Z = -21°$, β angle $= 105°$, perceptible $r > v$ dispersion and pleochroism $n_\alpha =$ pale green, $n_\beta =$ olive green, $n_\gamma =$ dark green or $Z > Y > X$.

This example is optically negative, since n_β is nearer n_γ, and has a $2V$ angle of about 56° (Fig. 8-9). The optic plane XZ is parallel to the {010} crystallographic plane, since $Y = b$ and b is normal to {010}, and by the construction shown in Figure 8-21,B, we see that $a \wedge X = -6°$ and that the angle between a and the nearest optic axis is 22°.

A section parallel to {010} (Fig. 8-21,B) will show the following:

1. Extinction angle of 21° (i.e., $c \wedge Z$) to the cleavage, which is the maximum extinction angle for sections through c.

2. Positive elongation, since the vibration direction of the slow wave (i.e., Z) lies nearer the cleavage.

3. Maximum birefringence $(n_\gamma = n_\alpha) = 0.054$, since X and Z lie in the plane of the section. In a thin section of standard thickness, this section should appear third-order red.

4. Refractive index $n_\gamma = 1.688$ and dark green when cleavage traces are 21° to the N–S cross hair (extinction position).

5. Refractive index $n_\alpha = 1.634$ and pale green when cleavage is 21° to the E–W cross hair (extinction position).

6. Flash interference figure, since Y is vertical.

A section parallel to {100} (Fig. 8-21,C) will show the following:

1. Parallel extinction, since optic plane and optic normal (Y) lie parallel and normal, respectively, to cleavage traces.

2. Refractive index $n_\beta = 1.676$ and olive green when cleavage is E–W, since Y lies in {100} normal to cleavage traces.

3. Refractive index between n_γ and n_β and color between dark green and olive green when cleavage is N–S, since the normal to {100} (i.e., the line of sight) lies between X (n_γ) and an optic axis (n_β) in the optic plane.

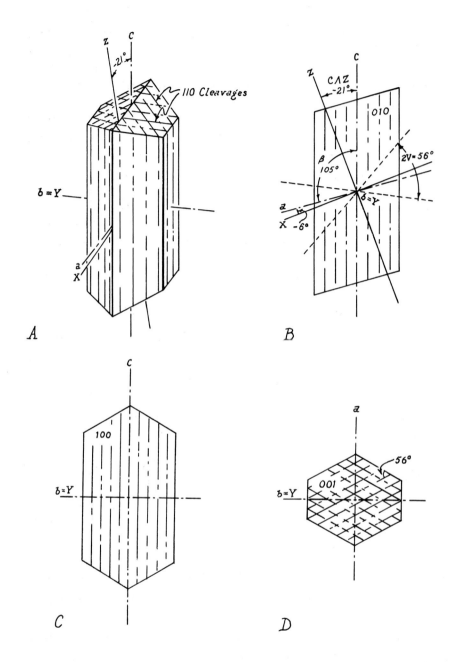

FIGURE 8-21. Optical orientation of hornblende. (A) Hornblende is monoclinic with $b = Y$, $c \wedge Z = -21°$, $2V = 56°$, $\angle \beta = 105°$, and perfect {110} cleavages. B, C, and D show crystal sections parallel to {010}, {100}, and {001}, respectively.

4. Positive elongation, since the wave vibrating parallel to the cleavage (i.e., in the optic plane) is the slow one (i.e., greater refractive index).

5. Low birefringence, which must be less than $(n_\gamma - n_\beta) = 0.012$, and, in standard thin section, retardation not exceeding first-order yellow.

6. Optic-axis interference figure 7° off center (i.e., an acute bisectrix 21° off center) showing perceptible $r > v$ optic-axis dispersion and inclined bisectrix dispersion. The normal to $\{100\}$ (i.e., the line of sight) lies in the optic plane at an angle of 7° from an optic axis and 21° from the acute bisectrix (X). And $b = Y$, producing inclined bisectrix dispersion in the acute bisectrix figure.

A section parallel to $\{001\}$ (Fig. 8-21,D) will show the following:

1. Symmetrical extinction to the two prismatic cleavages because the optic plane and optic normal bisect cleavage angles.

2. Refractive index n_β or 1.676 and olive green when the N–S cross hair bisects the acute cleavage angle, since Y lies in $\{001\}$ and bisects the acute cleavage angle.

3. Refractive index between n_β and n_α (nearer to n_α) and color between olive green and pale green when the N–S cross hair bisects the obtuse cleavage angle, since the normal to $\{001\}$ (i.e., the line of sight) lies in the optic plane between Z and an optic axis (nearer to Z).

4. No measurable elongation, since two intersecting cleavages do not cause general lineation. Note, however, that the slow wave vibrates parallel to the long dimension of the diamond pattern produced by intersecting cleavages.

5. Moderate birefringence, which must be somewhat less than $(n_\beta - n_\alpha) = 0.042$, and, in a standard thin section, retardation not exceeding third-order green.

6. Obtuse bisectrix interference figure 6° off center, which theoretically shows inclined bisectrix dispersion. The $\{001\}$ normal lies in the optic plane 6° from Z.

Cleavage fragments of hornblende tend to lie on $\{110\}$ cleavage planes; those that do may be expected to show:

1. Inclined extinction less than the maximum 21°.

2. One refractive index between n_γ and n_β and color between dark green and olive green when the fragment is in extinction with cleavage traces nearer the N–S cross hair.

3. A second refractive index between n_β and n_α and color between olive

green and pale green when the fragment is in extinction with cleavage traces closer to the E–W cross hair than to the N–S cross hair.

4. Positive elongation, since the slow ray vibrates close to the cleavage traces.

5. Moderate birefringence, obviously less than $(n_\gamma - n_\alpha)$ or 0.054.

6. An off-center interference figure, since all optical directions and optic axes lie at an appreciable angle from the $\{110\}$ normal.

Triclinic crystal forms are conventionally chosen so that prominent cleavages are pinacoidal. Cleavage possibilities in triclinic minerals, however, are essentially unlimited and unpredictable. Since optical orientation is also unlimited and unpredictable, no generalizations can be made about the relationship of specific cleavages to optical properties, except to note that sections or fragments showing parallel extinction, principal refraction indices, or centered interference figures are rare. Optical orientation of triclinic minerals is often described by stating that an optical direction or principal plane essentially parallels a crystallographic axis or crystal plane, in which case the optical properties of given sections may be predicted on the basis of the principles outlined in the preceding section and by constant reference to the appropriate indicatrix.

CHAPTER 9

The Universal Stage

Function and Construction

The universal stage of E. V. Fedorow was designed to allow mounted fragments or thin sections to be rotated to any orientation, an obvious advantage to anyone who has long engaged in the tedious search for a fragment or section accidentally oriented to give a particular interference figure or refractive index.

The Axes

The universal stage is bolted to the upper surface of the microscope stage where it allows thin sections or mounted fragments to be rotated about several axes in addition to the vertical axis of the microscope stage. Universal stages are manufactured with three, four, or five axes of rotation. Three axes are essential, a fourth is convenient for initial alignment, and the fifth axis is now virtually a historical curiosity. The "improved" or five-axis stage championed by S. F. Emmons serves for his double variation method of determining the refractive index, and the fifth axis allows some measurements to be made with less graphic plotting. It is very awkward to use, however, and the inexperienced student is encouraged to plot his data to clarify procedures and avoid errors. Only the four-axis stage will be considered in the pages to follow.

To facilitate description of procedures, it is necessary to name the axes of rotation. The following table presents the various systems in use:

	Emmons	Berek	Fedorow-Nikitin	Reinhard
Inner vertical axis (I.V.)		A_1	N	N
Inner east-west axis (I.E–W)		—	—	—
North-south axis (N–S)		A_2	H	H
Outer vertical axis (O.V.)		A_3	M	A
Outer east-west axis (O.E–W)		A_4	J	K
Microscope axis (M.)		A_5	—	M

The system of Fedorow and Nikitin has priority to recommend it; the system of Emmons makes use of six axes of rotation (including the microscope stage). The system of Berek, beginning with A_1 in the center and progressing outward to A_5, will be used in this work (Fig. 9-1).

Each rotation ring is marked in degrees for measuring angular rotation. Rotations about A_3 and A_4 are measured with vernier scales, and rotations about A_2 are measured on movable Wright arcs.

The Glass Hemispheres

Without glass hemispheres (Fig. 9-2) above and below the mineral section, light rays from below are highly refracted and reflected upon entering and leaving the section; consequently, the critical angle is reached with comparatively little rotation, and light does not enter the section at all. If the hemispheres, microscope slide, balsam, cover glass, mounting oil, and the mineral under observation all have the same index of refraction, light rays striking the lower hemisphere pass straight through the center of the entire unit to the objective, without refraction or deviation, regardless of rotation. This is, of course, a condition that is seldom, if ever, attained. Ideally, all layers between the hemispheres (i.e., glass slide, cover glass, balsam, oil, and mineral section) are bounded by perfectly smooth, parallel surfaces, and although the light path is bent at each surface, its direction in the upper hemisphere is parallel to that in the lower hemisphere. The observed ray direction is the light path through the glass of the hemispheres, which unfortunately is not the path of concern; we are concerned only with the direction of light within the mineral (Fig. 9-3). Since the refractive indices of hemisphere and mineral differ, the measured angle of rotation (i.e., $A_1 \wedge A_5$) must be corrected. Recall from page 11 that

$$\frac{n_2}{n_1} = \frac{\sin i}{\sin r}, \quad \sin i = \frac{n_2}{n_1} \sin r.$$

In this particular application, n_1 is the refractive index of the mineral, n_2 the index of the hemispheres, r the measured angle (between A_1 and A_5), and i the corrected angle (between A_1 and the light path through the mineral). (Since the refractive index of the mineral varies with direction, the appropriate index

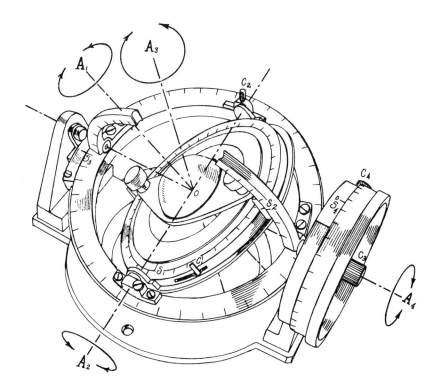

FIGURE 9-1. Rotation axes for a four-axis universal stage (E. Leitz Inc.). All rotation axes intersect at the center (O) of the flat side of the glass hemisphere. A_1 is the rotation axis normal to the plane of the thin section, which is placed between the glass hemispheres. Rotation about A_1 is read as S_1 on the periphery of the inner rotating ring. Clamp is at C_1. A_2 is the horizontal N–S axis when A_3 and A_4 are at zero. Rotation about A_2 is read at S_2 on the hinged Wright arcs. Clamp is at C_2. A_3 is normal to the A_2-A_4 plane, and rotation about A_3 is read S_3 on the periphery of the outer rotating ring. Clamp is at C_3. A_4 is horizontal and E–W when the microscope axis A_5 is at zero. Rotation about A_4 is read at S_4 on the large vertical drum on the right. Clamp is at C_4. A_5 (not shown) is the axis of the microscope stage.

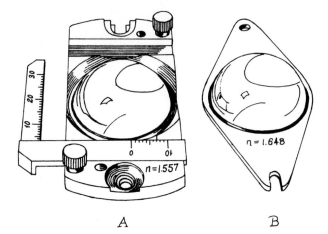

FIGURE 9-2. Upper hemisphere units (E. Leitz, Inc.). (A) An upper hemisphere with built-in Schmidt guide, which is used to control orientation of the thin section when orientation of several mineral grains within the section are to be related by universal stage measurements. The base of the glass slide rests against the numbered guide at the left. (B) The hemisphere used for most routine orthoscopic work is much less bulky and is easier to use.

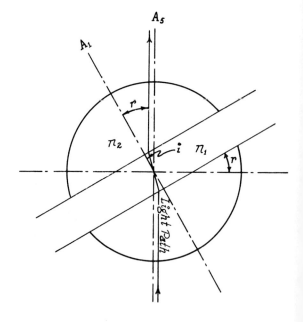

FIGURE 9-3. Light path through the spherical unit. Spherical surfaces prevent refraction of light rays between air and the glass hemispheres. Although there are many inclined surfaces in contact between the hemispheres (oil, glass, balsam, and mineral), as long as these surfaces are smooth and parallel they do not affect the final ray path. We observe the ray path in hemispheres (n_2) but wish to measure the ray path in the mineral (n_1). These paths are related by the expression $\sin i = (n_2/n_1) \sin r$, where i is the angle between A_1 and the light path in the mineral and r is the angle between A_1 and A_5.

of the given direction should be used. Unless the mineral has very high birefringence, the mean index of the mineral is usually sufficient.) The magnitude of the correction increases with the difference between the refractive indices of hemisphere and mineral and with inclination on A_2 or A_4. If $n_1 = n_2$, then $i = r$; or, if $r = 0°$, then $i = 0°$; in either case, there is no correction. The angle of inclination (i.e., the angle between A_1 and A_5) can theoretically vary from $0°$

to 90° but in practice seldom exceeds 50°. If the inclination does not exceed 40° and $n_2 - n_1$ does not exceed 0.10, the correction will not exceed 3°. Measurement errors of 2° or 3° are common, but if the difference in indices is large (e.g., > 0.10) or the inclination great (e.g., > 30°), the correction should be made and is easily read from Fedorow's diagram (Fig. 9-4).

In an attempt to reduce this correction, sets of hemispheres are available in a variety of different refraction indices so that a set may be chosen with an index that approximates that of the mineral under examination. An index is marked on every hemisphere pair.* Unfortunately, the surfaces between hemispheres are not exactly parallel, nor are they smooth (polished mineral surfaces are recommended). Consequently, accurate measurements are best obtained by using hemispheres and oil that have the same index as the glass slide, cover glass, and balsam (i.e., about 1.54) with appropriate corrections from Figure 9-4. Cedar oil or glycerin are traditional mounting oils for normal hemispheres ($n = 1.555$), and α-monobromonaphthalene may be used for high-index hemispheres ($n = 1.649$). If the mounting oil is mixed to exactly the same refractive index as the spheres, scratches on flat hemisphere surfaces will not show troublesome refractions.

The lower hemisphere is mounted in a metal ring in the center of the universal stage. Leitz, B & L, and Vickers upper hemispheres commonly bolt down on both sides of the stage, but Zeiss hemispheres bolt on one side only, and a screw head on the bolt is used to adjust the effective thickness of the slide, which should be between 1.40 and 1.42 mm. For petrofabric studies the frame of the upper hemisphere may be equipped with a Schmidt guide (Fig. 9-2,A) or one of similar design. Even a small mechanical stage may be clamped to some universal stages to allow the thin section to be moved across the field without changing orientation.

Upper and lower hemispheres of standard size have a radius of 13.5 mm. For conoscopic studies, upper hemispheres of smaller radius (7 mm and 3.5 mm) are used with standard lower hemispheres.

Special Objective Lenses

The greatest difficulty encountered in the development of the universal stage was not within the stage itself, but in the manufacture of satisfactory objective lenses to be used with it. Most high-power, or even medium-power, objectives have a small working distance (i.e., the distance between cover glass and lens

---

*E. Leitz manufactures hemisphere sets with refraction indices of 1.516, 1.554, and 1.684; C. Zeiss, 1.555 and 1.649; and B & L, 1.516, 1.559, and 1.649.

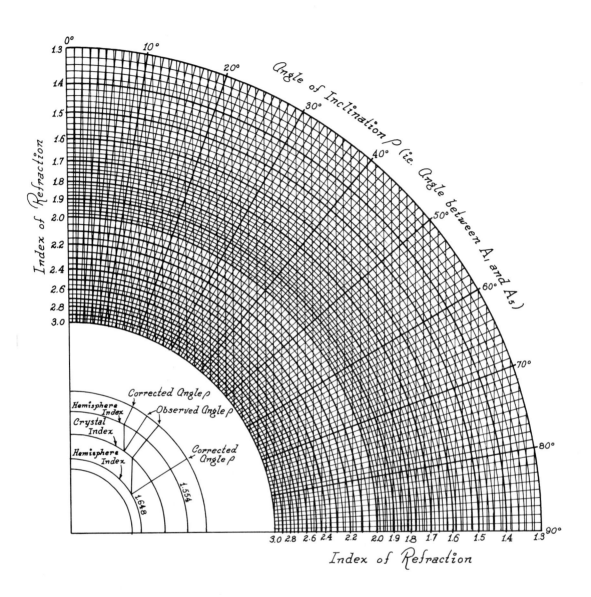

FIGURE 9-4. Fedorow's diagram for correcting the angle of inclination (ρ). The key to the use of this diagram is shown at the lower left corner. The radial line denoting the observed ρ (i.e., the angle between A_1 and A_5) is traced from the ρ scale inward toward the lower left until it intersects the arc that matches the index of the crystal. From this intersection a vertical line is followed either upward or downward until it intersects the arc that represents the refractive index of the spheres. The corrected ρ is found by tracing the radial line from the latter intersection outward and returning to the ρ scale. [After Emmons, 1943.]

surface), and most glass hemispheres have a radius of about 13.5 mm. This large working distance decreases the numerical aperture and, hence, the resolving power of the lens. Leitz produces special, large-element, objectives with long working distance (UM 5× to 30×) for use with standard hemispheres, and Zeiss manufactures low-power objectives (UD 6.3× and 16×) for use with standard hemispheres and high-power objectives (UD 20× and 40×) for use with small hemispheres.

Most objectives designed for use with the universal stage contain an iris diaphragm (Leitz) or separate diaphragm cones (Zeiss) to control the effective aperture within the lens.

Illumination

ORTHOSCOPIC ILLUMINATION. Light rays passing through the thin section should be parallel, but are converged by refraction at the hemisphere surfaces (especially when small hemispheres are used); consequently, only the central rays are nearly parallel. For normal orthoscopic observation, diaphragms below the polarizer and in the objective are closed down to exclude convergent rays, large hemispheres are used with low-power objectives, and the condenser system may be completely removed. Narrow diaphragms require a very intense light source for sharp extinctions. Early workers used shielded arc lamps. Illumination should be well centered (see p. 42).

CONOSCOPIC ILLUMINATION. Light rays passing through the mineral section converge to form interference figures. For conoscopic observation, diaphragms may be left open. Small hemispheres are used with high-power objectives (N.A. about 0.65), and a special condenser lens (N.A. about 0.6) is positioned just below the lower hemisphere to produce a strongly convergent light cone.

Mounting and Adjusting the Universal Stage

The universal stage is so constructed that all axes of rotation intersect at a common point when it is properly mounted and adjusted.

1. Thin sections or mounts are most easily placed between the hemispheres before the universal stage is attached to the microscope stage. When the Zeiss universal stage is used the upper surface of the lower hemisphere acts as base for the thin section; when the Leitz stage is used a separate glass plate, set in the center ring of the universal stage, acts as base for the section. One or two

drops of oil are put on the base to lubricate the surface for the thin section. Another drop is put on top of the section, and the upper hemisphere is clamped in place above the section with appropriate mounting screws. The section should glide easily on the films of oil when moved with the fingertips. When the Leitz stage is used, a rotation of 180° about A_2 exposes the bottom of the glass plate; after adding a drop of oil, the lower hemisphere is pressed against the plate.

2. The center plate of the microscope stage is removed to allow greater rotation of A_2 and A_4, and the universal stage is bolted to the upper surface of the microscope stage with the rotation drum for the A_4 axis facing right (see Fig. 9-1).

3. The objective lens is focused on the section and centered in the conventional manner (p. 43) on A_5.

4. With A_2 and A_4 at zero, A_1 is made to coincide with A_5. When A_1 is rotated, one point in the field of view remains stationary while the field rotates about it. This stationary point is brought to the center of the field by using the centering screws on the base of the universal stage or, if no centering screws are present, by loosening the stage bolts, sliding the complete universal stage on the microscope stage, and bolting it securely in place.

5. If the plane of the thin section does not lie in the A_2-A_4 plane, the section rotates about an axis above or below its plane, and the field of view moves either north or south as A_4 is rotated. Several projections on the underside of the central metal ring allow the threaded ring to be rotated to raise or lower the section. The threads are conventional right-hand threads and should be turned accordingly to raise or lower the section. It may be necessary to loosen the upper hemisphere to raise the section. If the field moves in the same direction as A_4 is turned, the section needs to be lowered until the field remains stationary about A_4, and conversely.

6. The final adjustment is to align A_2 and A_4 with the north-south and east-west cross hairs. The objective lens can be focused on specks of dust on the top or bottom of the glass sphere. As A_4 is rotated, the dust particles should move north and south parallel to the cross hair. If this motion is not parallel, the microscope stage (A_5) is rotated until it is, A_5 is clamped, and this position becomes the zero position for A_5, from which all rotations of this axis are measured. (This new zero position will never be more than one or two degrees from the zero mark on the periphery of the stage.) Rotation about A_2 should move dust specks parallel to the E–W cross hairs, and slight rotation on A_3 may be necessary to assure this alignment.

Graphical Representation of Measurements

Three-dimensional data may be represented on a two-dimensional plane surface by several simple procedures, as one may recall from other studies in geology (e.g., crystallography and structural geology).

The Spherical Projection

Each of these procedures begins with a sphere with the phenomena to be considered at its center. Any plane or line within the sphere is represented by its *pole*, which is the point of intersection of the line or plane normal with the spherical surface. The spherical projection is a spherical surface containing poles of planes and lines which has little practical value except as the initial step in the development of the plane projections (Fig. 9-5).

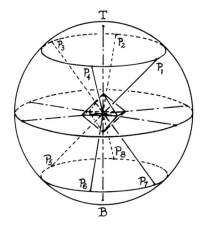

FIGURE 9-5. Spherical projection. Lines radiating from the central point represent optic axes, crystallographic directions, twin axes, normals to cleavages or crystal faces, and so on. The imaginary spherical surface containing the points of intersection of each line (i.e., pole) is the spherical projection of these phenomena.

THE GNOMONIC PROJECTION. This plane projection is formed by an observer at the center of the sphere (C in Fig. 9-6). His line of sight projects the pole P_5 to point G on the projection plane. Obviously, poles in the upper hemisphere cannot be projected to this surface; and for values of angle ρ near 90°, the extent of the projection plane approaches infinity.

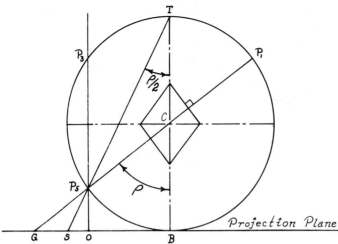

FIGURE 9-6. Plane projections. Plane projections are formed by the projection of a spherical projection to a plane surface. Gnomonic projection (G) of P_5 is formed by the line from the center of the sphere (C) through P_5. Orthographic projection (O) of P_5 is formed by the line from infinity on TB through P_5. Stereographic projection (S) of P_5 is formed by the line from T through P_5.

THE ORTHOGRAPHIC PROJECTION. For this plane projection, the observer is located at infinity on the vertical axis (line TB) so his line of sight is parallel to the vertical axis. The orthographic projection of pole P_5 is at O. All points of this projection lie within a circle of unit radius. Poles in the upper hemisphere may project at the same point as poles in the lower hemisphere (P_3 and P_5).

THE STEREOGRAPHIC PROJECTION. Here the observer is located at the top (T), or bottom, of the spherical surface. His line of sight projects the pole P_5 to point S on the projection plane. The elevation of the projection plane affects only the relative size of the total projection, and, the stereographic projection plane is often drawn through the center of the sphere. All poles in the lower hemisphere project within a unit circle.

The Wulff Net

If a sphere is marked by the longitude-latitude method, turned on its side, and stereographically projected to a plane surface, the resulting pattern is a Wulff net* (Fig .9-7). With this net, one can plot the position of any stereographically projected pole, without actually making the projection, and measure the angle

---

*Printed Wulff nets may be obtained from Ward's Natural Science Establishment, Rochester, N.Y. (10 cm and 14 cm diameter), the U.S. Navy Hydrographic Office (40 cm dia.).

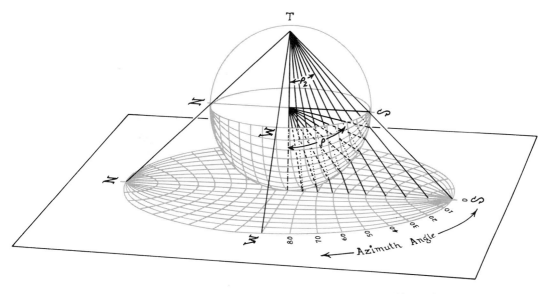

FIGURE 9-7. Wulff net projection. The Wulff net, or stereographic net, is a stereographic projection of the longitude and latitude lines on a sphere.

between any two lines from their projected poles. When the statistical distribution of poles is important (e.g., as in petrofabric studies), an equal-area, or Schmidt, net is used. A given area anywhere on the Schmidt net represents an equal area anywhere on the spherical projection (Fig. 9-8).

PLOTTING THE POSITION OF A LINE (DIRECTION) ON A WULFF NET (*Figs. 9-9 and 9-10*). The Wulff net is usually mounted on a thin board, and a pin or small nail is driven from the back side through the exact center of the net. Points are not plotted directly on the net but on a tracing paper overlay impaled on the pin so as to rotate about the center point of the net. The tracing paper represents the stereographic projection plane, and the points plotted on it represent the projected poles when the thin section is in a horizontal position (i.e., when A_2 and A_4 are at zero).

The line CP_5 in Figure 9-6 might represent the direction of an optic axis, a crystallographic axis (*a, b, c*), an optical direction (*X, Y, Z*), a twinning axis, a cleavage normal, the normal to a twinning or composition plane, or any other direction of interest to the investigator. By a few simple rotations, to be described later, the direction representing one of the above is oriented either vertical, parallel to A_5, or horizontal, parallel to A_4. To locate the projection

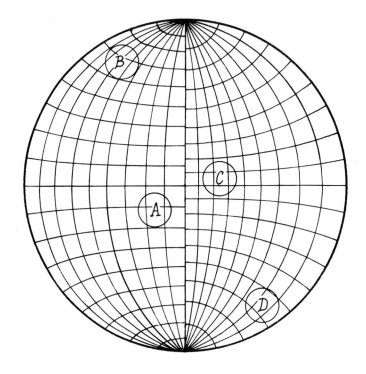

FIGURE 9-8. Comparison of Wulff (stereographic) and Schmidt (equal-area) nets. The equal-area net (left side) is constructed so that equal areas anywhere on the surface of the sphere remain equal on the projection (e.g., areas *A* and *B*). It is used in petrofabric studies where one may wish to evaluate the statistical distribution of poles over the surface of the spherical projection. If ten poles appear in one square centimeter on the spherical surface and five in an equal area anywhere else on the sphere, the appropriate areas of equal size on the equal-area projection will also contain ten and five poles respectively. The stereographic net (right side) is distorted in such a way that equal areas on the projection (e.g., areas *C* and *D*) do not represent equal areas on the surface of the sphere. Note that an area 10° longitude $\times$ 10° latitude near the poles is much smaller than an area 10° longitude $\times$ 10° latitude near the equator. This is true not only on the spherical surface but on both kinds of projection.

of this line on the projection plane (tracing paper), two angles of rotation are required (see Fig. 9-7): (1) the azimuth angle, measured on the periphery of the unit circle, and (2) the angle of inclination to the projection plane ρ.

1. The azimuth angle is the angle which the thin section is turned from zero about A_1. To represent this angle on the projection plane (Fig. 9-9), a small arrow or zero mark, is drawn on the tracing paper at "0°" on the periphery of the net,* and the tracing paper is rotated about the center pin, the same angle the thin section is rotated about A_1 (i.e., azimuth angle).

2. The ρ angle is the angle between the line (direction) to be plotted and the normal to the projection plane. The projection plane is normally parallel to the plane of the thin section and, since A_1 is always normal to the thin section, it represents the normal to the plane of projection. The line to be plotted is oriented parallel to either A_5 or A_4, and ρ becomes the angle between A_1 and either A_5 or A_4. With A_2 and A_4 at zero, A_1 is parallel to A_5 and perpendicular to A_4, and ρ is either 0° or 90°. If the line to be plotted (e.g., optic axis) is parallel to A_1 ($\rho = 0°$), the projection is at the exact center of the net (regardless of the azimuth angle). If the line to be plotted is parallel to A_4 ($\rho = 90°$), the projection falls on the periphery of the net at both ends of the east-west line, when the azimuth angle is appropriately set. Some rotation of either A_2 or A_4 is ordinarily required to align the desired direction with A_5 or A_4, and the ρ angle can be read directly from the Wright arc used to measure rotations about A_2 or from the drum used to measure rotations about A_4.

If the axis, or other desired direction, is located parallel to A_5 (Fig. 9-9,A), the ρ angle is scaled off, on the net, from the center outward along either the east-west line, if the rotation is on A_2 (Fig. 9-9,B); or on the north-south line, if the rotation is on A_4. To determine which direction to move from the center, one should observe the direction of movement of the pole (in the lower hemisphere) as A_2 or A_4 is returned to its zero position (Fig. 9-9,B and 9-10,B). Sometimes rotations on both A_2 and A_4 are necessary to align a certain direction with A_5. In this case, the appropriate angle of rotation of A_2 should be measured from the center outward along the east-west diameter and the angle of rotation of A_4 is scaled off to the north or south along the great circle which intersects the east-west diameter at the point determined by the A_2 rotation.

If the axis or other desired direction is oriented parallel to A_4 (Fig. 9-10,A), the ρ angle is measured from the periphery of the net inward toward the center along the east-west diameter, for A_2 rotations (Fig. 9-10,B).

---

*When the universal stage is properly mounted and all axes at their zero position, the zero on the A_1 scale is in the "south position."

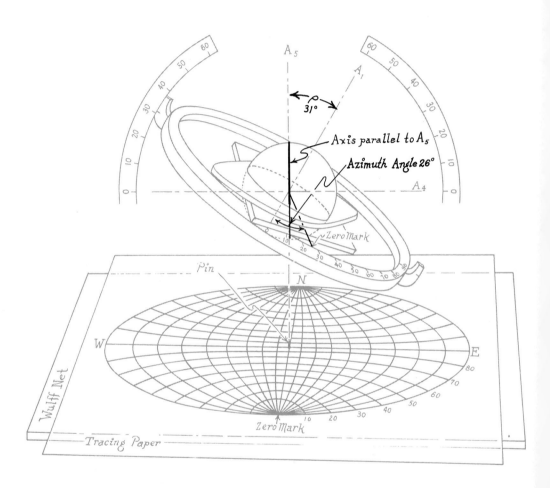

FIGURE 9-9. Plotting an axis oriented parallel to A_5. The heavy line on A_5 in Part A of the figure represents an optic axis that has been oriented parallel to A_5 by a rotation of 26° about A_1 (azimuth angle) and of 31° about A_2 (ρ), as described in the text. This direction (axis) plots at S (part B) by rotating the tracing paper 26° about the center of the net and

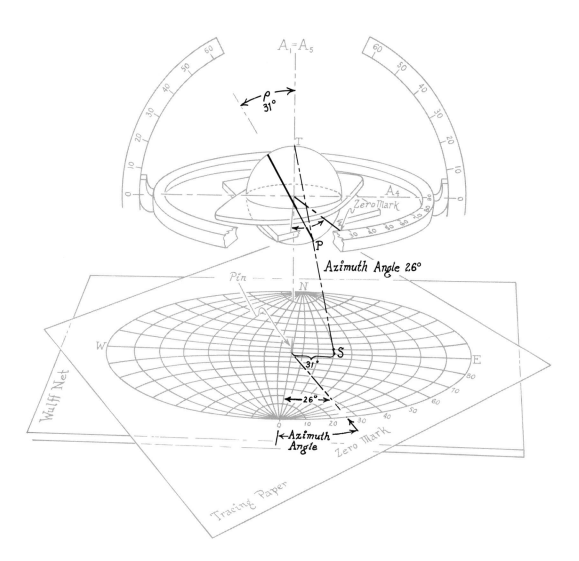

moving along the E–W line to 31° east of the center. If orientation of the axis were achieved by rotation about A_4 instead of A_2, then S would lie on the N–S line, either north or south of center.

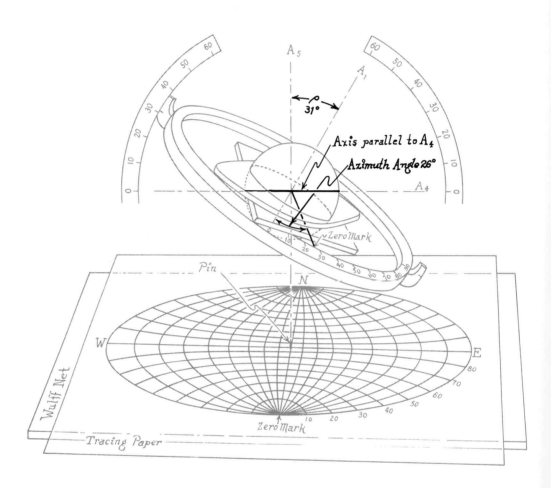

FIGURE 9-10. Plotting an axis oriented parallel to A_4. The heavy horizontal line on A_4 in part A of the figure represents an optic axis, crystallographic direction, cleavage normal, or some other line oriented parallel to A_4 by rotation of 26° on A_1 (azimuth angle) and 31° on

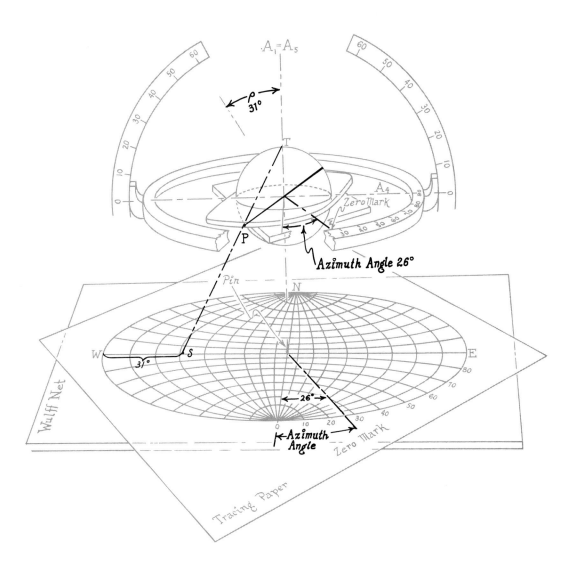

$A_2(\rho)$ as described in the text. This direction plots at S (part B) by rotating the tracing paper 26° about the center of the net and moving in 31° from W on the E–W line.

188   *The Universal Stage*

PLOTTING THE TRACE OF A PLANE (*Fig. 9-11*). The projected trace of any plane can be drawn by plotting the position of its pole, rotating the paper until the pole lies on the east-west line, scaling of 90° along this line and drawing the great circle which passes through this point.

FIGURE 9-11. Plotting the trace of a plane. (Below) The plane is oriented vertically and N–S so that its normal parallels A_4. The normal is then plotted as shown in Figure 9-10. (Right) The trace of the plane is on the great circle 90° from the projection of its normal.

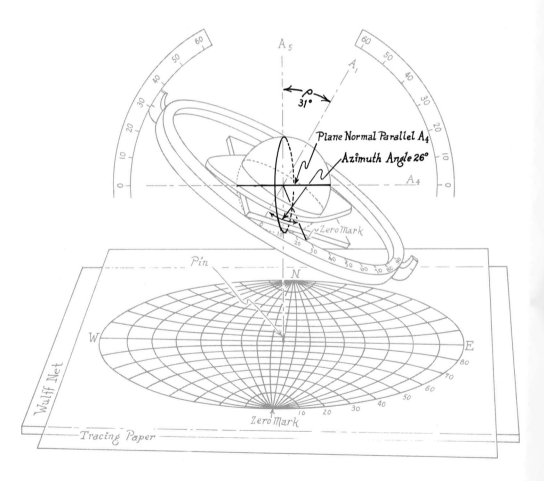

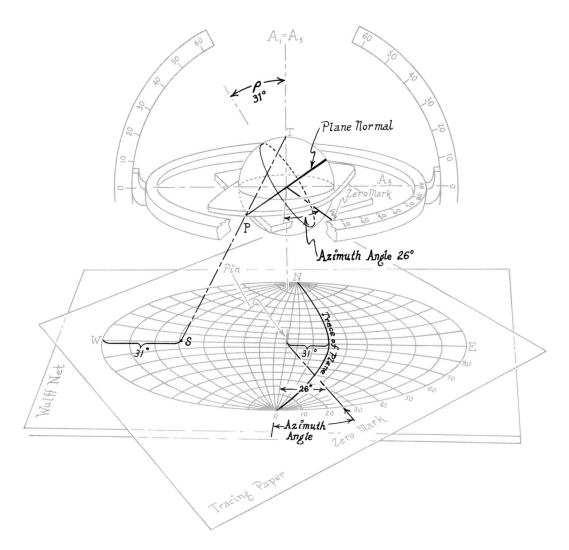

ROTATING POINTS ON THE PROJECTION PLANE (*Fig. 9-12*). It is frequently desirable to rotate the entire projection so that some convenient direction is either vertical or horizontal. To graphically effect this rotation, the overlay paper is rotated until the projected point, representing the line to be made vertical, lies on the east-west line and the point is moved the necessary number of degrees along this line to the center of the net. Without moving the paper, all other points are moved along the small circles, on which they lie, the same number of degrees and in the same direction as the first point. If a point is rotated "off the edge of the net" the point representing the projection of the "other end of the line" will enter the net 180° (azimuth angle) from where the first point left, and the rotation is completed from there.

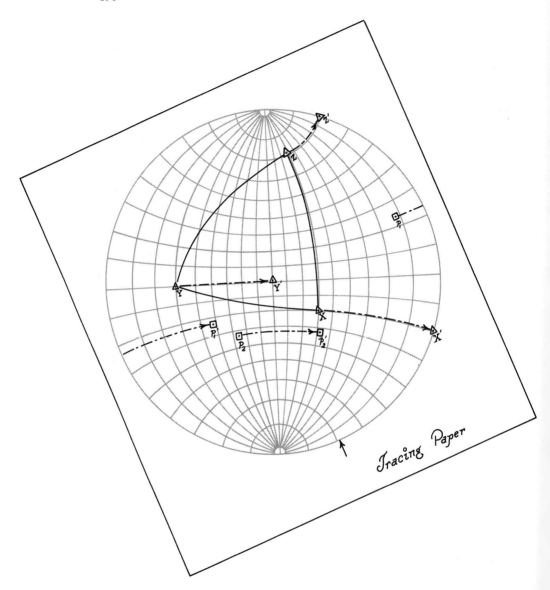

FIGURE 9-12. Rotating points to change orientation of the stereogram. This figure is a plot of a biaxial crystal, showing the orientation of X, Y, and Z (90° apart) and two points P_1 and P_2 which may represent cleavage normals, twin axes, etc. It may prove advantageous to rotate the entire stereogram such that Y is vertical and the XZ-plane horizontal. To do this, Y is rotated to the E–W line and moved 60° east to Y' at the center. Every other point on the stereogram is rotated 60° eastward along its small circle. Note that P_1 rotates off the stereogram to reappear 180° around the periphery of the circle and complete its 60° rotation to P_1'.

CHAPTER 10

Application of the Universal Stage to Uniaxial Crystals

Distinguishing Uniaxial and Biaxial Minerals

In every section of a uniaxial crystal, vibration directions bear a simple relationship to the optic axis. The ordinary wave vibrates perpendicular to it, and the extraordinary wave in a plane containing it. In random sections of biaxial crystals, however, the vibration directions show no obvious relationship to optical directions X, Y, and Z.

With all axes at zero, a birefringent mineral grain rotated about A_1 will extinguish when its inherent vibration directions lie N–S and E–W. Since the optic axis of a uniaxial mineral must lie in either the N–S or E–W vertical plane at any extinction position, the mineral grain will remain extinct when rotated about either A_2 or about A_4,* whichever parallels the vibration direction of the ordinary wave (Fig. 10-1,A,B,C).

At any extinction position about A_1, the extinction of a biaxial mineral is relieved by rotation about both A_2 and A_4 separately (Fig. 10-1,D,E,F), excluding special orientations in which an optical direction, X, Y, or Z, lies parallel to either A_2 or A_4. One should examine several grains to avoid special orientations.

---

*The mineral grain will remain extinct about both A_2 and A_4 if the optic axis lies in the plane of the slide.

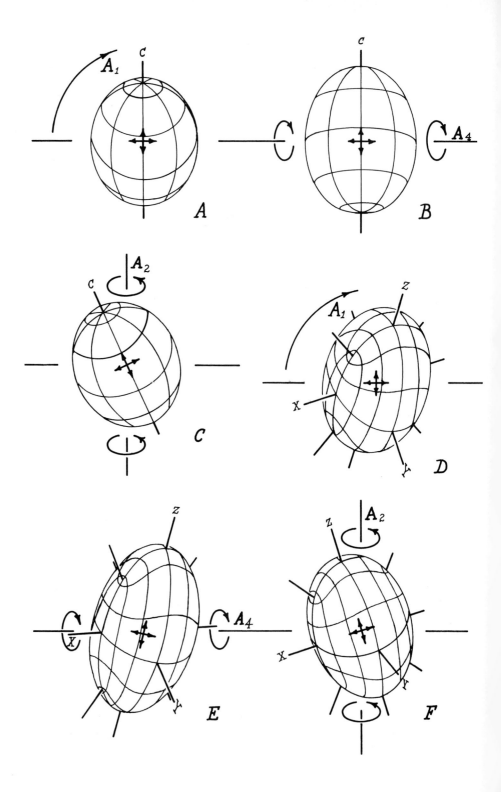

FIGURE 10-1. Distinguishing uniaxial and biaxial crystals. (A) A uniaxial crystal section in random orientation is represented by its indicatrix. Rotation about A_1 produces extinction when vibration directions (short perpendicular arrows) of waves passing through the center of the indicatrix parallel vibration directions of the nicols. The vibration direction of the ordinary wave is always perpendicular to the optic axis (i.e., c). (B) Rotation about the vibration direction of the ordinary wave (A_4 in this example) does not relieve extinction. (C) Rotation about the vibration direction of the extraordinary wave (A_2 in this example) does relieve extinction, unless the optic axis lies in the plane of the section. (D) A biaxial crystal section, in random orientation, is represented by its indicatrix. Rotation about A_1 produces extinction when vibration directions (short heavy arrows) of waves passing through the center of the indicatrix parallel vibration directions of the microscope nicols. (E) Rotation about A_4 relieves extinction unless X, Y, or Z parallels A_4. (F) Rotation about A_2 also relieves extinction unless X, Y, or Z parallels A_2.

Orientation of Uniaxial Crystals

Locating the Uniaxial Optic Axis by Orthoscopic Illumination

Under crossed nicols, rotation of a selected uniaxial mineral grain to extinction about A_1* places its optic axis in either the N–S or E–W vertical plane, and extinction will persist about either A_2 or A_4. If the grain remains dark when rotated about A_2, the optic axis is in the E–W vertical plane; if the grain becomes light about A_2 and remains extinct about A_4, the optic axis is in the N–S vertical plane. The optic axis is oriented either vertical, parallel to A_5, or horizontal, parallel to A_4, as follows (Fig. 10-2):

1. If the optic axis is not already in the E–W vertical plane, it is so oriented by a 90° rotation about A_1 to the next extinction position.

2. The crystal becomes light when turned forward 20° or 30° on A_4, excluding the special case in which the optic axis lies in the plane of the section (see the following page).

3. Extinction is restored by rotation about A_2. Depending on the direction of rotation about A_2, the optic axis will be either horizontal, parallel to A_4, or vertical, parallel to A_5, when A_4 is returned to zero. In either case the crystal remains extinct for rotation about A_4. With A_4 at zero, rotation about the microscope stage (A_5) indicates that the optic axis is vertical,† parallel

---

*A mineral grain that remains dark on rotation about A_1 is either isotropic or is oriented with an optic axis vertical. Successive rotations about A_2 and A_4 will always relieve the extinction of an anisotropic crystal.

†A mineral section of low birefringence is suggestive of a near-vertical optic axis and its optic axis is usually manuevered parallel to A_5. Conversely, high birefringent sections lend themselves to the orientation where c parallels A_4.

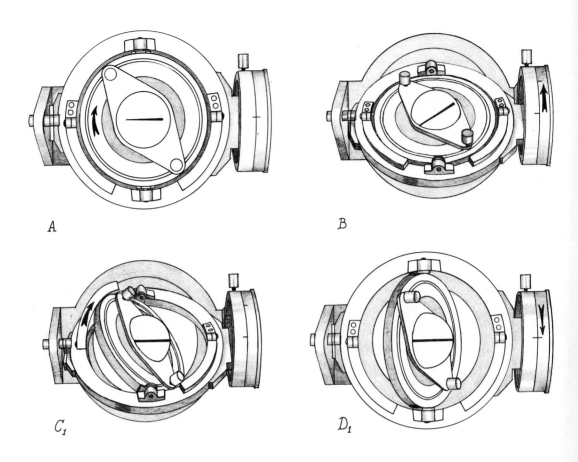

FIGURE 10-2. Locating the uniaxial optic axis. (A) Rotation to extinction about A_1 places the optic axis in either the E–W or N–S vertical, plane. (B) Rotation about A_4 relieves extinction if the optic axis is in the E–W plane (if not, rotate A_1 90°). (C) Rotation to extinction on A_2 places the optic axis either parallel to A_4 (C_1) or in the N–S vertical plane (C_2). In either

to A_5, if the crystal remains extinct. It is horizontal, parallel to A_4, if the crystal becomes light by rotation about A_5.

The orientation of the optic axis can now be stereographically plotted from an azimuth angle (i.e., rotation about A_1) and a ρ angle (i.e., rotation about A_2), the latter being measured along the E–W diameter from the center outward if the optic axis is vertical or from the periphery inward if the optic axis is horizontal.

If the optic axis lies in the plane of the slide, a crystal section at extinction

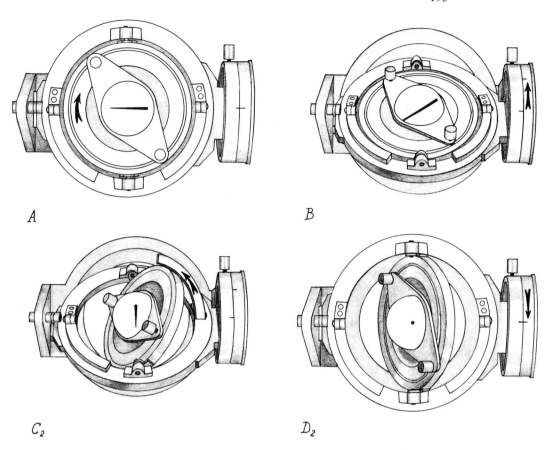

case, extinction persists about A_4. (D) When A_4 is returned to zero, the optic axis is either horizontal and parallel to A_4 (D_1) or vertical and parallel to A_5 (D_2). Rotation about A_5 indicates a vertical axis if extinction persists.

---

about A_1 remains dark about both A_2 and A_4, and the optic axis parallels either A_2 or A_4. If the optic axis parallels A_4, the crystal becomes birefringent when rotated some arbitrary amount about A_4 and then about A_2.

THE OPTIC SIGN. With the optic axis parallel to A_4, a 45° counterclockwise rotation about A_5 places the optic axis parallel to the direction of slow vibration for accessory plates. Addition of retardation indicates that the extraordinary wave is the slower of the two, and hence that the mineral is positive; conversely, subtraction indicates that the mineral is negative.

Locating the Uniaxial Optic Axis by Concoscopic Illumination

Conoscopic illumination allows common interference figures to be seen and a melatope to be rotated to the center of the field (i.e., the optic axis can be rotated into a vertical position). Because orthoscopic extinction is usually incomplete or complete over a wide range, conoscopic illumination allows a melatope to be oriented with greater precision.

Without the Bertrand lens, a uniaxial mineral grain is rotated to extinction on A_1; when the lens is inserted, an isogyre can be seen to cut the field either N–S or E–W. Slight adjustment about A_1 centers the isogyre on a cross hair, and rotation about A_2 (E–W isogyre) or A_4 (N–S isogyre) allows us to follow the isogyre to an optic axis (indicated by a centered optic-axis figure) or to an optic normal (indicated by a flash figure). If the interference figure is sharp, a vertical optic axis can be oriented and plotted within an accuracy of 0.5°. Optic sign is determined by accessory plates in the usual manner.

Cleavage Studies

Cleavage type can be determined by representing the c-crystallographic axis (i.e., the optic axis) and the traces of cleavage planes on a stereographic plot (stereogram). A cleavage normal is plotted by rotating the cleavage plane to the N–S vertical plane, in which position the cleavage normal is parallel to A_4. The mineral section is rotated about A_1 until the cleavage plane is N–S and then about A_2 until it is vertical. Cleavage traces are wide and diffuse when the cleavage is inclined, but become thin and sharp as the cleavage becomes vertical.* A white line, similar to a Becke line, may appear on both sides of a vertical cleavage trace, moving to one side as the cleavage is inclined. When the cleavage normal is parallel to A_4, it may be plotted on the net in the same manner as an optic axis parallel to A_4. To draw the cleavage trace on the stereographic projection, start with the cleavage normal on the E–W line of the net, mark off 90° along the E–W line, and draw the great circle passing through this point to represent the cleavage trace.

The relative positions of cleavage trace, cleavage normal, and c-axis distinguish three basic types of cleavage (Fig. 10-3).

1. *Basal cleavage* is characterized by a single cleavage direction that becomes apparent when the cleavage normal coincides with the c-axis. The mineral

---

*An error of two or three degrees is to be expected.

grain is, of course, in the same position for plotting both cleavage normal and c-axis.

2. *Prismatic cleavages* are indicated when cleavage normals are 90° to the c-axis. If the c-axis is rotated until it lies on the E–W diameter of the net, the normals of all prismatic cleavages fall on the arc 90° to c. Prismatic cleavages of tetragonal minerals occur in pairs (i.e., (100) and (010) or (110) and ($1\bar{1}0$)), and the cleavage normals lie 90° apart along the arc. If both sets of prismatic cleavages are present, the four cleavage normals show 45° separation. Prismatic cleavages of hexagonal minerals occur in sets of three and intersect at angles of 120°, 60°, or 30°.

3. *Pyramidal cleavages* are relatively uncommon except for trigonal minerals that show rhombohedral cleavage. Pyramidal cleavage is indicated when the c-axis falls neither on cleavage normal nor on cleavage trace. It occurs in sets of four or eight for tetragonal minerals, in sets of six or twelve for hexagonal minerals, and in sets of three or six for trigonal minerals.

Twinning Studies

A *simple twin* consists of two individuals (i.e., either two complete crystals or two crystal halves) geometrically united to appear as though one individual were formed from the other by reflection across a plane, by rotation of 180° about an axis, or, more commonly, by both. The plane of reflection, which becomes a symmetry plane by virtue of the twinning, is called the *twinning plane*; the axis of rotation is called the *twinning axis*; and the plane along which the two individuals are united is the *composition plane*, and is usually coincident with the twinning plane. Simple twins may be either *contact twins* (Fig. 10-4,A), two crystal halves united by a composition plane, or *penetration twins* (Fig. 10-4,B), two complete crystals that appear to interpenetrate one another. *Repeated twinning*, in which more than two individuals are joined, may be either *polysynthetic* (Fig. 10-4,C), repeated contact twinning, or *symmetrical* (cyclic) (Fig. 10-4,D), repeated penetration twinning.

A twin is visible in thin section between crossed nicols because optical orientation is different for the united individuals. Except in positions of special orientation, interference colors differ on either side of the composition plane.

The composition plane of a twin is located and graphically represented in the same way as a cleavage plane, by plotting the normal to the plane. When the composition plane of a normal twin* is vertical, interference colors on both sides of the plane are the same.

---

*A normal twin is one in which the twin axis is normal to the composition plane (twin plane).

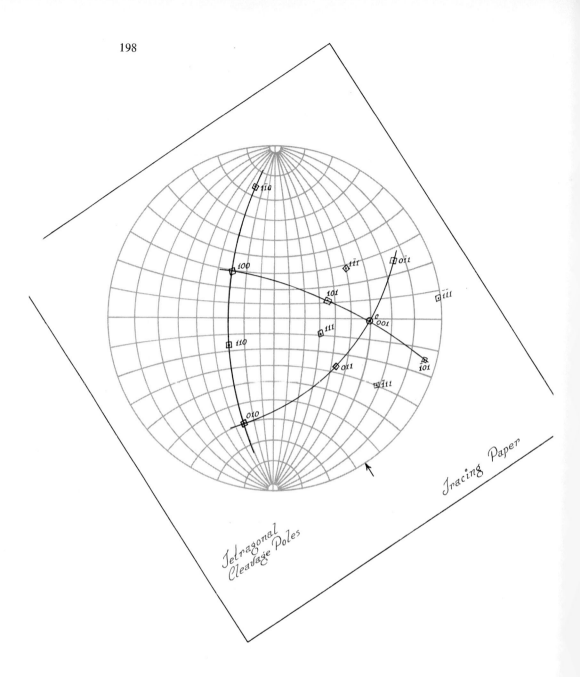

FIGURE 10-3. Cleavage poles of uniaxial crystals. Cleavages in uniaxial crystals must be basal, {001} or {0001} with one direction; prismatic, {hk0} or {hki0} with two or three directions; or pyramidal, {hkl} or (hkil) with three, four, or six directions. The pole of basal cleavage coincides with the *c*-crystallographic (optic) axis. Poles of prismatic cleavages must lie in the basal plane (i.e., on the trace of the plane normal to the optic axis). Prismatic cleavage poles of tetragonal crystals lie in the basal plane at multiples of 45°, and those of hexagonal crystals at multiples of

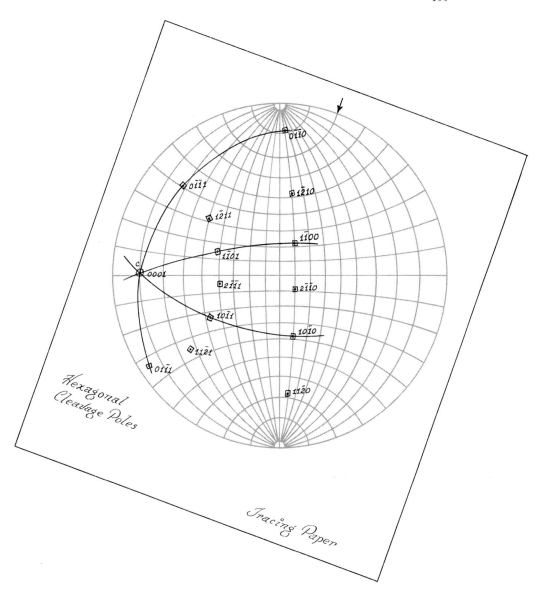

30°. Poles of pyramidal cleavages coincide neither with the optic axis nor with the basal plane. Pyramidal cleavage poles of tetragonal crystals lie on planes through c and intersect the basal plane at multiples of 45°; those of hexagonal crystals lie on planes through c and intersect the basal plane at multiples of 30°.

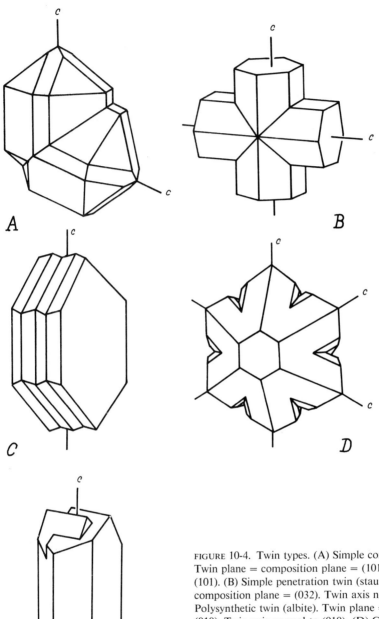

FIGURE 10-4. Twin types. (A) Simple contact twin (cassiterite). Twin plane = composition plane = (101). Twin axis normal to (101). (B) Simple penetration twin (staurolite). Twin plane = composition plane = (032). Twin axis normal to (032). (C) Polysynthetic twin (albite). Twin plane = composition plane = (010). Twin axis normal to (010). (D) Cyclic twin (crysoberyl). Twin plane = composition plane = (031). Twin axis normal to (031). (E) Parallel penetration twin (sanidine). No twin plane. Composition plane = (010). Twin axis parallel to c.

By plotting the position of the c-axis on both sides of the composition plane—the trace of the composition plane and the twin axis (i.e., the normal to the composition plane)—the twinning axis and twinning plane may be defined with respect to the c-axis (Fig. 10-5).

Since an axis of even symmetry cannot serve as a twinning axis, and since planes of symmetry cannot be twinning planes, the possible variety of twins in the tetragonal, trigonal, and hexagonal crystal systems is severely limited by high symmetry. For uniaxial minerals, only normal twins are possible, except in those rare crystal classes that contain either no center of symmetry or no plane of symmetry.*

By far the most common tetragonal twins occur in minerals of the di-tetragonal-dipyramidal class (e.g., rutile, cassiterite, zircon, idocrase), in which the pyramidal plane, $\{011\}$ or $\{101\}$, is the twinning plane and the twin axis is normal to the twinning plane. Twinning is commonly repeated as cyclic twinning, and the measured angle between twinning axis and c-axis reveals the three-fold (twin axis $\wedge$ $c \simeq 60°$), five fold (twin axis $\wedge$ $c \simeq 35°$), or eight-fold (twin axis $\wedge$ $c \simeq 22°$) nature of the cyclic twinning.

Twins are unusually rare and unimportant in the hexagonal system, but many twins exist in the trigonal system. Very common are normal twins of the trigonal carbonates, especially calcite, where the twinning plane is either the basal pinacoid $\{0001\}$ or a pyramidal plane as positive rhombohedron $\{10\overline{1}1\}$ or negative rhombohedron $\{01\overline{1}2\}$. Alpha quartz belongs to the rare trigonal trapezohedral class, which has neither a plane nor a center of symmetry. It may twin by reflection only with $\{1\overline{2}10\}$ as the twin plane (Brazil law), by rotation only with the c-axis as twin axis (Dauphine law), or by both rotation and reflection with a pyramidal plane $\{11\overline{2}2\}$ as twinning plane and a normal twin axis.

Distinguishing Carbonate Minerals in Thin Section

A simple method of distinguishing calcite, dolomite, and magnesite in thin section (described by R. C. Emmons in G.S.A. Memoir 8) takes advantage of high birefringence and the refraction index of balsam (1.54), which lies between n_ϵ and n_ω for these minerals:

A carbonate grain is selected that is in contact with balsam, preferably one

---

*Crystals that have no center of symmetry may twin by rotation only, yielding twins with twinning axes but no twinning planes; crystals that possess no plane of symmetry may twin by reflection only, forming twins with twinning planes but no twinning axes.

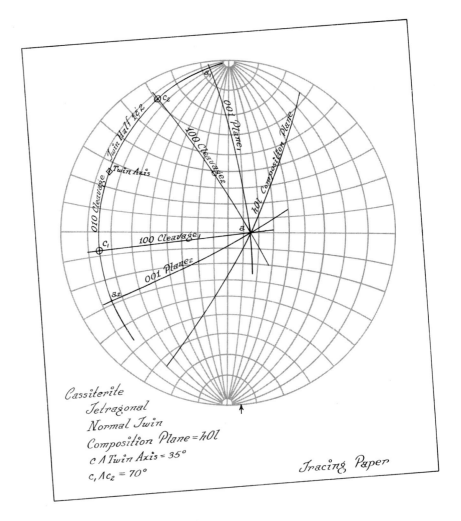

whose high birefringence indicates that its optic axis lies near the plane of the section. The grain is moved to the center of the field, and the optic axis is located by orienting it horizontal parallel to A_4 as described on page 194. Rotation of 90° about A_3 moves the optic axis into a N–S horizontal position. With nicols uncrossed, the carbonate grain shows negative relief, since its index of refraction is n_ϵ ($n_\epsilon = 1.486$ for calcite, $n_\epsilon = 1.500$–1.526 for dolomite, and $n_\epsilon = 1.509$–1.527 for magnesite). Rotation about A_4 raises the optic axis toward the vertical, and the refractive index is raised from n_ϵ, which

FIGURE 10-5. Uniaxial twinning. Cassiterite, the example used here, is a tetragonal mineral with imperfect {100} cleavages (i.e., parallel to 100 and 010). It commonly forms twins of the kind shown in Figure 10-4,A. By plotting c for both halves of the twin, the composition plane, and cleavages {100} and {010}, we obtain the following information: (1) Since cassiterite is uniaxial, any twin must be normal (i.e., the twin axis is normal to the composition plane = twin plane). (2) Axes c and one a must lie on the trace of each cleavage, {100} and {010}. The c-axis on each twin half must lie at the intersection of cleavages {100} and {010} on that half of the twin. (3) One prismatic cleavage, here labeled {010}, is common to both twin halves, and its trace passes through both c-axes (c_1 and c_2) and must contain two a axes (a_1 and a_2) 90° to each c. (4) The other prismatic cleavage, {100}, has different orientation on the two halves of the twin. The trace of each {100} cleavage passes through one c and intersects the composition plane at a common point 90° to c. This intersection must be an a-axis common to both crystal halves. (5) The trace of the basal plane {001} on each twin half is represented by a great circle through both a-axes on that crystal half. (6) Because the trace of the composition plane passes through only one a-axis and neither c-axis, it is a pyramidal (h01) or 0k1) plane. (7) The twin axis lies in the plane defined by the two c-axes about 35° from each c, and the twin plane is about 55° to basal planes {001}.

is below 1.54, toward n_ω, which is above 1.54; at some angle of rotation about A_4, the refractive index of the mineral grain equals 1.54 and its relief falls to zero.

The angle through which A_4 must be rotated from zero on the A_4 scale to zero relief should be about 36° for calcite, 29° for dolomite, and 25° for magnesite. Since it is difficult to know when the optic axis is exactly horizontal, it is more accurate to measure the angle between the two positions of zero relief by turning A_4 forward to zero relief and then backward to zero relief. Since isomorphous substitution of calcium and magnesium, and even other elements, is extensive in the common carbonates, a range of values seems more appropriate than the single values given above. The most practical values for the angles between the two directions of zero relief appear to be:

Calcite 72°–64°

Dolomite 60°–52°

Magnesite 52°–38°

It is desirable to determine this angle for several separate grains and calculate an average.

CHAPTER 11

Application of the Universal Stage to Biaxial Minerals

Orientation of Biaxial Minerals

Locating Optical Directions (X, Y, and Z) with Orthoscopic Illumination

Consider vibration directions indicated by a biaxial indicatrix (Fig. 10-1), and note that (1) vibration directions of light waves traveling in a random direction have no readily apparent relationship to the optical directions X, Y, and Z, and (2) as the indicatrix is rotated about a horizontal E–W axis (i.e., A_4), vibration directions remain N–S and E–W only when one optical direction X, Y, or Z parallels the rotation axis and the plane of the other two is vertical N–S.

The optical directions are located and identified by placing each direction (X, Y, and Z) successively parallel to A_4, all observations being made between crossed nicols as follows:

1. The biaxial mineral grain under examination is rotated about A_1 to extinction. The vibration directions of the mineral grain now parallel the cross hairs, but nothing is certain about the location of X, Y, or Z.

2. Tipping A_4 forward, or backward, relieves extinction, unless one of the optical directions parallels A_4.

3. With the slide still rotated some arbitrary amount on A_4, it is tipped on A_2 until extinction is restored.

4. When A_4 is returned to zero the mineral grain is again bright but less bright than before (i.e., nearer to extinction).

5. A slight rotation about A_1 again restores extinction.

6. Repetition of 2, 3, 4, and 5 brings some optical direction ever nearer A_4, and when the mineral grain remains extinct about A_4, either X, Y, or Z parallels the rotation axis A_4 and the plane of the other two optical directions is vertical N–S.

This optical direction is plotted on the stereographic (Wulff) net by reading the azimuth angle on the A_1 scale and the ρ angle on A_2 (Fig. 9-10). With the point representing this optical direction still on the E–W line, 90° is marked off from the point along the E–W line, and the great circle is drawn to represent the vertical N–S plane on which the other two optical directions must fall (Fig. 11-1).

With an optical direction located and plotted, it must now be identified as X, Y, or Z. The first step is to rotate the entire universal stage 45° counterclockwise about A_5 so that the unknown optical direction is horizontal NE–SW and the mineral grain is bright again.

If the plotted optical direction is Y, the optic plane (XZ) is vertical NW–SE, and within that plane lie two optic axes (Fig. 11-1). Rotation about A_4 may bring one or both optic axes into coincidence with A_5 when the birefringence of the mineral grain falls to zero. Rotation about A_5 confirms a vertical optic axis, since the mineral grain remains dark. The position of this optic axis (both optic axes if possible) is recorded on the stereogram; keep in mind that the optic axis must lie on the great circle that represents the XZ-plane at an angle above or below the E–W line as indicated on the A_4 scale. Occasionally both optic axes of a mineral with small $2V$ will lie near the plane of the section, and it may prove mechanically impossible to align an optic axis with A_5. In that case, however, the interference colors descend the scale as A_4 is rotated toward its maximum rotation, indicating that an optic axis is approaching the vertical.

If the plotted optical direction is either X or Z, no position of extinction can result from rotation about A_4 in the 45° position about A_5. Use of the gypsum plate, or other accessory, reveals whether the wave vibrating parallel to A_4 is fast or slow. If the wave parallel to A_4 is the fast wave (i.e., interference colors subtract) the optical direction parallel to A_4 is X. If the wave proves to be slow, the optical direction Z parallels A_4.

To locate a second optical direction, A_2, A_4, and A_5 are returned to zero, and the mineral grain is rotated about A_1 to its other extinction position. Repetition of the steps above brings a second optical direction into coincidence with A_4. (If this procedure should result in the optical direction that was initially located again being parallel to A_4, return to the starting position and repeat all steps, rotating A_4 in the opposite direction to that in step 2.) This optical direction is identified as X, Y, or Z, and its position recorded on the stereogram, where it must lie on the trace of the plane 90° to the first optical direction. An error of two or three degrees is common, and a repeated measurement of both optical directions will frequently reduce or completely eliminate the error.

The third optical direction lies at the intersection of the planes 90° to each of the other two (Fig. 11-1), and its identity is obtained by elimination. Where mechanically possible, it is desirable to adjust the universal stage so that the third optical direction parallels A_4 in order to check its exact position and confirms its identity. This is done by moving the point representing the third optical direction to the E–W line of the stereogram and reading the A_1 setting from the periphery and the A_2 setting along the E–W line.

THE 2*V* ANGLE AND OPTIC SIGN. With Y on the E–W line of the stereogram, the angle between an optic axis and the nearest optical direction, measured along the XZ great circle, is the V angle, and $2V$ is twice this value. If Z is nearer an optic axis than X, then Z is the acute bisectrix and the mineral is optically positive. If X is nearer, it is the acute bisectrix, and the mineral is negative.

With Y parallel to A_4 in the 45° position about A_5, the $2V$ angle can often be measured directly by rotation about A_4 from one extinction position (i.e., one optic-axis vertical) to the second (i.e., the other optic-axis vertical).

Locating Optical Directions (X, Y, and Z) by Conoscopic Illumination

With consoscopic illumination, common interference figures are seen and optical directions may be rotated to required orientations with greater precision than is possible with the uncertain extinctions attained with orthoscopic illumination.

1. Without the Bertrand lens, a biaxial mineral grain is rotated to extinction about A_1; when the lens is inserted, an isogyre becomes visible, commonly cutting the field as an arc in oblique orientation.

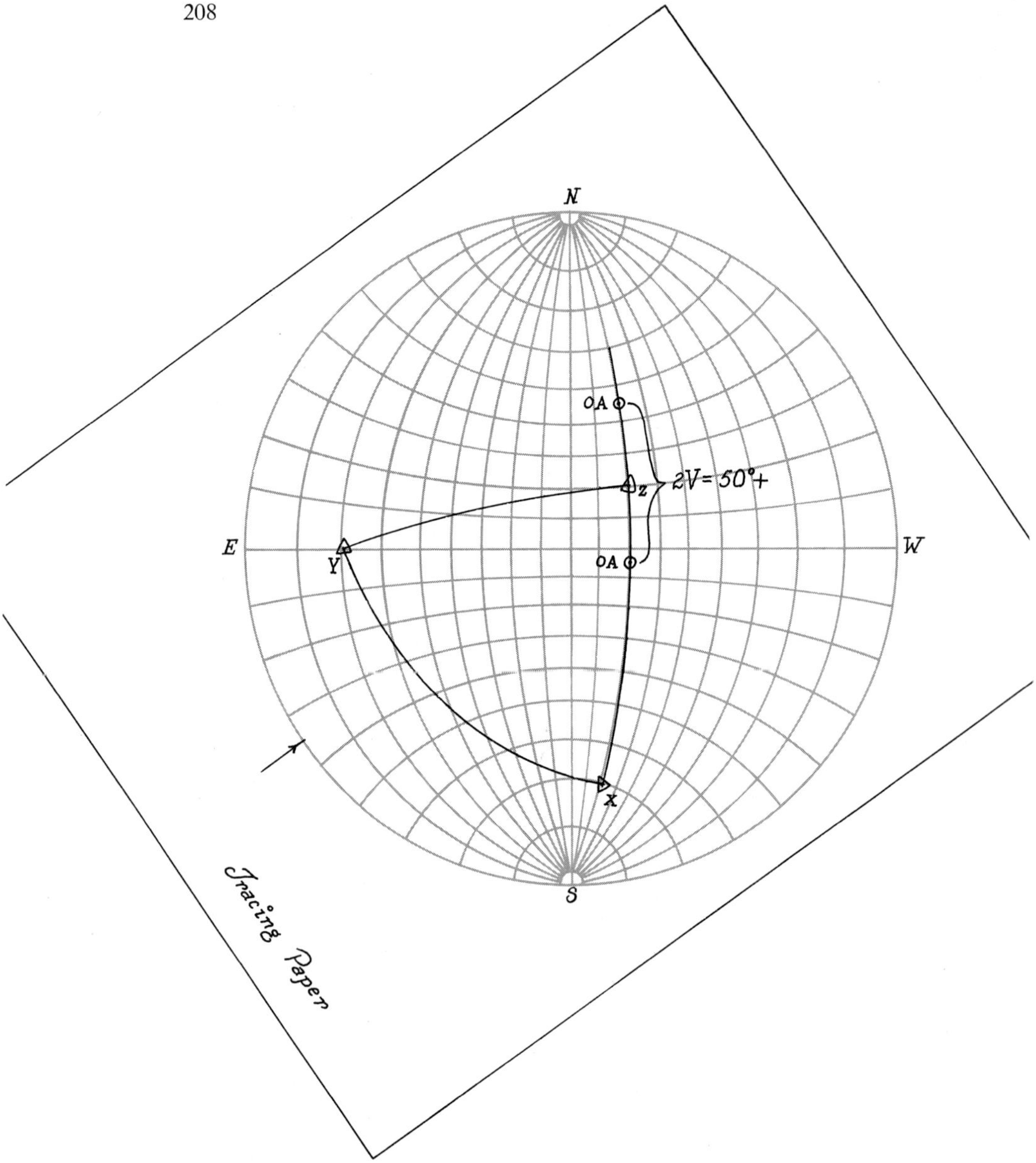

FIGURE 11-1. Stereogram of a biaxial crystal. The optical direction X was located parallel to A_4, as described in the text, and plotted on the E–W line; the great circle 90° to X was drawn to represent the YZ-plane. The optical direction Y was then located and the stereogram rotated to plot Y on the E–W line as shown. The great circle 90° to Y is the XZ-plane, and Z lies at the intersection of the YZ- and XZ-planes. With Y parallel to A_4, one or both optic axes may be located, as described, and plotted on the XZ (optic) plane. Since Z is the acute bisectrix, this crystal is optically positive; the $2V$ angle measures about 50°.

2. Minor rotation on A_2 positions the arc so that it intersects the center of the field.

3. Slight rotation on A_1 straightens the isogyre and aligns it as nearly N–S as possible.

4. Repetition of 2 and 3 produces a straight N–S isogyre. One principal plane is now vertical N–S, and one optical direction (X, Y, or Z) parallels A_4.

5. Rotation on A_4 enables us to follow the isogyre, which is bisected always by the N–S cross hair, to the acute bisectrix, obtuse bisectrix, or optic normal, as indicated by the appropriate, centered bisectrix interference figure or flash figure. The interference figure is in its 90° position (i.e., straight isogyres intersect at the center of the field), and the isogyre intersection at an acute bisectrix can often be aligned with the cross hair intersection within 0.5°.

6. One optical direction (X, Y, or Z) is now vertical, parallel to A_5, another parallels A_4, and the third is horizontal N–S. All three directions can be plotted on the stereogram.

Rotation about A_5 may allow the vertical optical direction to be recognized as the acute bisectrix, obtuse bisectrix, or optical normal. If so, the optic sign can be determined in the usual way with an accessory plate, and the optical directions can be identified by their relation to the orientation of the interference figure. The acute bisectrix is either Z (positive) or X (negative), and the obtuse bisectrix is the other of the two; the optical normal (Y) is perpendicular to the optic plane (i.e., the trace through the two melatopes). The vertical optical direction may be difficult to recognize, especially if $2V$ is large. Where mechanically possible, however, each optical direction may be brought successively to the vertical position for comparison, using its known position on the stereogram.

The 2V Angle and Optic Sign. When a bisectrix is vertical and the optic plane is N–S, 45° rotation on A_5 separates the isogyres to their 45° position, with the optic plane diagonally bisecting the field. Rotation about A_4 (i.e., Y), first forward then backward, brings both melatopes successively to the center of the field, and the $2V$ angle is measured directly by the A_4 rotation between melatopes. If the refractive index of the glass spheres is nearly the same as that for balsam (i.e., 1.54), if correction is made for refraction (Fig. 9-4), if the interference figure is clear and sharp, and if $2V$ is measured at both 45° positions about A_5, the average $2V$ should be correct within $\pm 0.5°$.

Optic sign is determined with accessory plates in the usual way.

Note that dispersion (optic axis and bisectrix) can be observed in any orientation on both isogyres, together or separately.

Relationships Between Optical and Crystallographic Directions

The procedure for determining the orientation of optical directions X, Y, and Z in a biaxial mineral grain is outlined above. Locating the crystallographic directions a, b, and c in the mineral grain requires the presence of some known cleavage, a twinning plane, a crystal boundary, or some other visible feature bearing a known relationship to one or more crystallographic axes. For example, both a and c must lie on the trace of a {010} cleavage, a {010} twinning plane, or a {010} crystal face, and c must fall on the trace of {110} cleavage. The stereographic projection of c must then lie at the intersection of {010} and {110} planes or at the intersection of any two prismatic planes.

 In orthorhombic crystals, optical directions parallel crystallographic directions. If one knows or can guess the Miller indices of observed cleavages, twinning planes, and crystal faces, he can often locate crystallographic directions at the intersections of these planes on the stereographic projection and can match the optical directions to the appropriate crystallographic directions to position the indicatrix. Cordierite, for example, shows cleavage parallel to {010} and commonly twins on {110} (Fig. 11-2). By plotting X, Y, and Z for half of the twin and the trace of the cleavage and twin plane, one observes that Y and X fall on the trace of the cleavage {010} and that X falls on the trace of the twin plane {110}, where it intersects the cleavage. Since c must parallel both cleavage and twin plane, c must fall at their intersection, which means that $c = X$. Since a must also parallel the cleavage {010}, we know that $a = Y$ and, by elimination, that $b = Z$.

 In monoclinic crystals, one optical direction parallels the b-crystallographic direction. Again, knowledge of cleavage, twin planes, or crystal faces is necessary to locate crystallographic directions. Orthoclase, for example, shows perfect cleavage parallel to {001}, less perfect parallel to {010}, and imperfect parallel to {110} (Fig. 11-3). Although four cleavage directions are required, only two or three directions are commonly visible in a given orthoclase grain. Orthoclase also commonly twins by the Carlsbad law, in which {010} is the composition plane. When X, Y, and Z are plotted for one side of a simple Carlsbad twin, it is found that both X and Y fall on the trace of the composition plane {010}. Since b is perpendicular to {010}, we know that $b = Z$. The trace of one cleavage parallels the {010} trace in thin section, but its trace on the Wulff net misses X, Y, and Z. This cleavage must be {110}, since {001} would pass through $b = Z$. The intersection of {010} and {110} is c, which lies 15° from Y in the $XY = ac$ plane, and $c \wedge Y = 15°$. By rotation about A_4, when the prismatic cleavages are approximately N–S, one may see the {001}

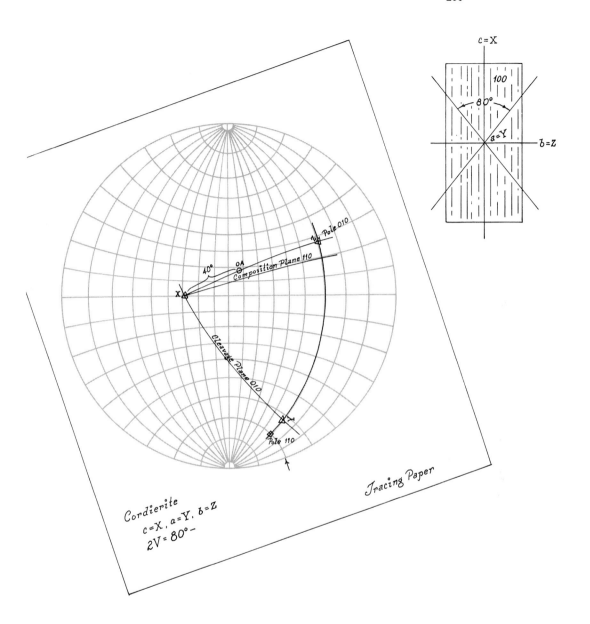

FIGURE 11-2. Orientation of cordierite. The stereogram shows orientation of X, Y, and Z and the trace of {110} twin composition plane and {010} cleavage. Since c parallels (i.e., lies on the trace of) both composition plane and cleavage, c must parallel X. Since a must lie on the trace of the {010} cleavage 90° from c, then a must parallel Y and, by elimination, b must parallel Z.

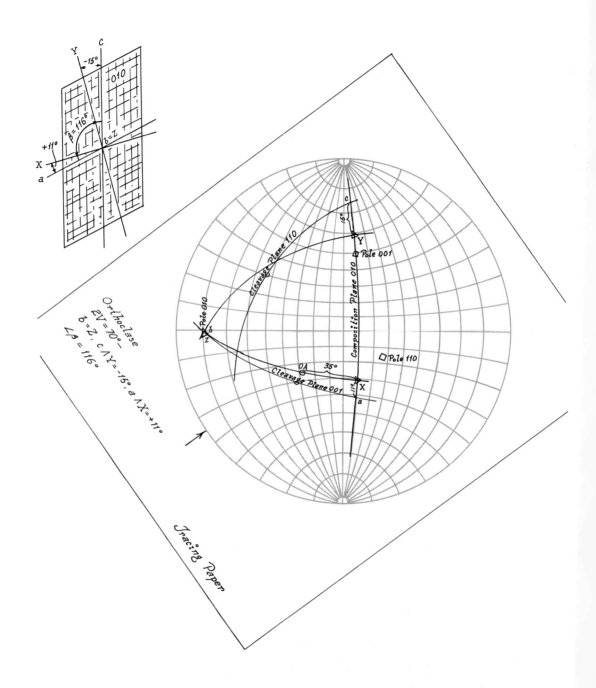

FIGURE 11-3. Orientation of orthoclase. Locating X, Y, and Z for one half of a twinned orthoclase crystal, the trace of the twin composition plane, which is known to be {010} (Carlsbad twin), and the trace of two cleavages enables us to describe the optical orientation of orthoclase. Since b is normal to {010} and Z is the pole of the composition plane, $b = Z$. We know orthoclase has {001}, {010}, and {110} cleavage. Since the cleavage plane passing through b must be {001}, the cleavage trace, which does not pass through b and is not normal to it, has to be {110}. The intersection of {010} and {110} must be c, and that of {010} and {001} must be a. When measured from c toward the obtuse β angle, $c \wedge Y = -15°$, and $a \wedge X = +11°$ when measured from a in the opposite direction. The β angle ($a \wedge c$) equals $90° + 15° + 11° = 116°$.

cleavage and plot its position. The trace of {001} passes through $b = Z$, intersecting {010} at a. By placing $b = Z$ on the E–W line and measuring along the $XY = ac$ great circle, one may find that $\angle \beta$ (i.e., $a \wedge c$) $= 116°$, $c \wedge Y = -15°$,* and $a \wedge X = +11°$.

Amphiboles and pyroxenes are mostly monoclinic with $b = Y$ and two perfect cleavages that are parallel to (110) and (1̄10) and intersect at about 87° and 93° (pyroxenes) or 56° and 124° (amphiboles). Using a typical cross section showing both prismatic cleavages,† one may locate X, Y, Z and the optic axes and plot the cleavage traces of both prismatic cleavages (Fig. 11-4). If measurements are accurate, cleavage traces will cross on the plane $XZ = ac$, which marks the position of c. Commonly a small triangle of error is present, and c is marked on the XZ-plane halfway between the two points at which the cleavage traces intersect XZ. The angle between c and Z (i.e., $c \wedge Z$) can be measured along the XZ-plane.** Although the pyroxenes and amphiboles allow a great variety of isomorphous substitution and the varieties are not well defined, it is usually possible to identify the more common varieties using $c \wedge Z$ and the $2V$ angle. Color and mode of occurrence are often necessary to distinguish between two or three varieties rendered possible by $c \wedge Z$ and $2V$.

Triclinic minerals are characterized by noncoincidence of optical and crystallographic direction. Sometimes, however, the optical directions lie close to crystallographic directions, within the error of measurement. Ordinarily, insufficient knowledge is gleaned from cleavages, composition planes, and

---

*The angle between c and Y is negative because it is measured from c toward the obtuse angle (see p. 160).

†Theoretically, only one cleavage is necessary, since it must intersect the XZ-plane at c, but greater accuracy is possible when both cleavages are used.

**For orthorhombic pyroxenes and amphiboles $c \wedge Z = 0°$, and $c = Z$.

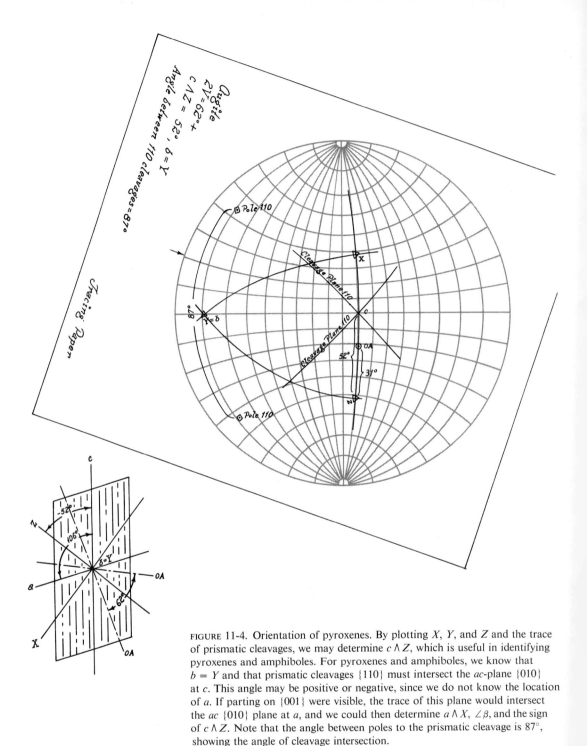

FIGURE 11-4. Orientation of pyroxenes. By plotting X, Y, and Z and the trace of prismatic cleavages, we may determine $c \wedge Z$, which is useful in identifying pyroxenes and amphiboles. For pyroxenes and amphiboles, we know that $b = Y$ and that prismatic cleavages {110} must intersect the ac-plane {010} at c. This angle may be positive or negative, since we do not know the location of a. If parting on {001} were visible, the trace of this plane would intersect the ac {010} plane at a, and we could then determine $a \wedge X$, $\angle \beta$, and the sign of $c \wedge Z$. Note that the angle between poles to the prismatic cleavage is 87°, showing the angle of cleavage intersection.

crystal faces to allow location of *a*, *b*, and *c*, since they bear no known relationship to X, Y, and Z. Occasionally, however, a twinned triclinic crystal showing several cleavages will yield sufficient data to locate *a*, *b*, and *c* and position the indicatrix (see the section on twinning below).

Cleavage Studies

If known cleavages can be used to determine crystallographic directions, knowledge of crystallographic directions should be useful for defining existing cleavages. Optical directions X, Y, and Z are located by optical means and represented as points on a stereogram. If their angular relationship to crystallographic directions, *a*, *b*, and *c* is known, these can also be plotted on the projection. The crystallographic directions which parallel a cleavage trace drawn on the stereogram indicate its Miller index. If *a* and *b* lie on the cleavage trace, the cleavage is {001}; if *a* and *c* fall on the trace, the cleavage is {010}; etc. If only *c* falls on the cleavage trace, the cleavage is prismatic (*hk*0); if no crystallographic direction parallels the cleavage, it is pyramidal (*hkl*).

Twinning Studies

Twinning in biaxial minerals is more complex and, consequently, more useful than twinning in uniaxial minerals. By plotting the optical directions X, Y, and Z for both individuals of the twin, the trace of the composition plane, and its normal, the student obtains a stereogram from which he may derive a wealth of information. He will also find it useful to plot cleavage normals and traces when cleavages are present.

Types of Twinning

Three distinct types of twins are possible in biaxial minerals, and they are defined by the relationship between twin axis and composition plane.

1. A *normal twin* exhibits a twin axis perpendicular to the composition plane. Nearly all common twinning laws are normal except a few important laws in the monoclinic and triclinic crystal systems.

2. A *parallel twin* is formed when the twin axis lies in the composition plane parallel to a crystallographic direction, *a*, *b*, or *c*.

3. A *complex twin* results when the twin axis lies in the composition plane perpendicular to a crystallographic direction, a, b, or c. A complex twin is, in effect, equivalent to a simultaneous combination of some normal twin and a parallel twin.

A normal twin is recognized when both individuals of the twin show the same interference color (crossed nicols) when the composition plane is vertical and the plane normal parallels A_4. Rotation about A_4 causes interference colors of both individuals to change together so that the twin is not apparent. This allows the composition plane of normal twins to be located with great accuracy.

A parallel twin or a complex twin, except under special conditions of orientation, shows different interference colors on either side of its composition plane when the plane is vertical N–S. Rotation about A_4 causes interference colors of both twin individuals to change independently. At some angle of rotation about A_4, however, both individuals assume the same interference color and the twin visually disappears. When this happens, the twin axis is either vertical and parallel to A_5 or horizontal and N–S. If rotation about the microscope stage (A_5) causes the two twin individuals to change colors together and extinguish at the same angle of rotation, the twin axis is vertical (parallel to A_5) and may be plotted on the composition plane above or below the E–W line as indicated by rotation from zero on A_4. If the interference colors change independently on the two individuals and they extinguish at different angles, the twin axis is horizontal N–S and may be plotted on the composition plane from the north or south pole as indicated on A_4. The plotted position of the twin axis for parallel and complex twins is usually not found with the same accuracy as for normal twins. If no optical direction is coincident for both twin individuals, the plotted position of the twin axis may be checked by mechanically rotating the stereogram until the X-optical directions for both twin individuals lie on the same great circle and then drawing the great circle. By repeating the process for both Y-directions and both Z-directions, the three great circles will be shown to intersect, forming a small triangle. The twin axis lies on the trace of the composition plane at the geometrical center of the triangle.

The twin can be identified as parallel or complex only if crystallographic directions can be located along the composition plane. Parallel twins are more common than complex twins, and if the trace of an unknown cleavage intersects the composition plane at the twinning axis, the twin is probably parallel and the intersection probably a, b, or c.

Twinning in Orthorhombic Minerals

Orthorhombic crystals form only normal twins, and the twin plane is usually {110}. Such a twin yields a stereographic pattern in which one optical direction X, Y, or Z coincides for both twin individuals and all other optical directions fall on the great circle at 90° (Fig. 11-5). The point of coincidence is the c-axis, and the trace of the composition plane passes through this point and intersects the 90° plane (i.e., the ab or {001} plane) halfway between points representing another optical direction of the two twin individuals and 90° to the twin axis, which also lies on the {001} plane. The angles between a or b axes for the two twin individuals may be measured along the {001} great circle.

Pinacoidal planes are impossible twin planes (they are symmetry planes), and all prismatic twin planes or planes parallel to domes produce twins in which one optical direction coincides for both twin halves; this direction parallels the crystallographic direction, which lies in the composition plane. Pyramidal (hkl) twin planes yield twins without coincidence of X, Y, or Z (Fig. 11-6).

Twinning in Monoclinic Minerals

Monoclinic crystals form normal or parallel twins in which the composition plane is usually {001} or {110} for normal twins and {010} for parallel twins. A normal twin with {001} or {110} composition plane produces a stereographic pattern like that of a common orthorhombic twin, with the coincidence of one optical direction and all other optical directions falling on the plane at 90° (Fig. 11-7). The coincident optical direction must parallel the b-crystallographic direction, and the composition plane passes through this point intersecting the ac or {010} plane at a point half way between two equivalent optical directions. This point of intersection is either the a crystallographic direction (001 composition plane) or the c-crystallographic direction (100 composition plane). Other composition planes are possible, but the stereographic patterns shows no coincidence of optical directions unless the composition plane is {$h0l$}.

Amphiboles and pyroxenes display perfect {110} cleavage (two directions), and $b = Y$. The composition plane of the common twin passes through both b and c, and is therefore {100} (Fig. 11-8).

The most common parallel twin in the monoclinic system forms according to the Carlsbad law, in which {010} is the composition plane and the twin axis parallels the c axis.

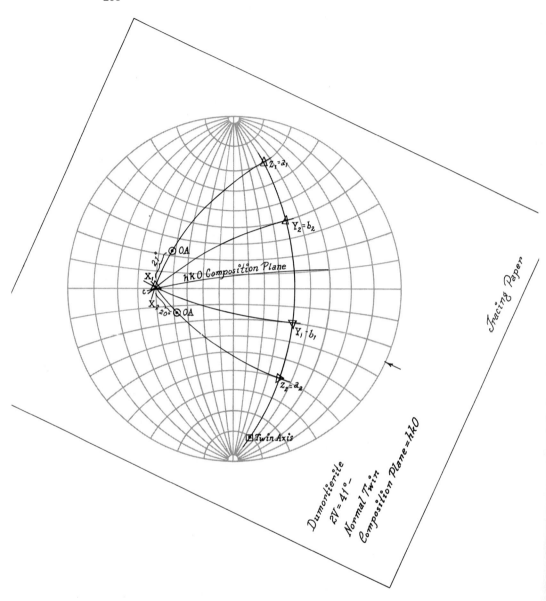

FIGURE 11-5. Dumortierite twin. We know dumortierite to be an orthorhombic mineral with $a = Z$, $b = Y$, and $c = X$. By plotting X, Y, and Z for both sides of the twin and the composition plane, we find the trace of the composition plane to pass through c only, making the composition plane $\{hk0\}$, probably $\{110\}$. We know the pole of the composition plane to be the twin axis and the twin to be normal, because both sides of the twin show equal birefringence, when the composition plane is vertical N–S.

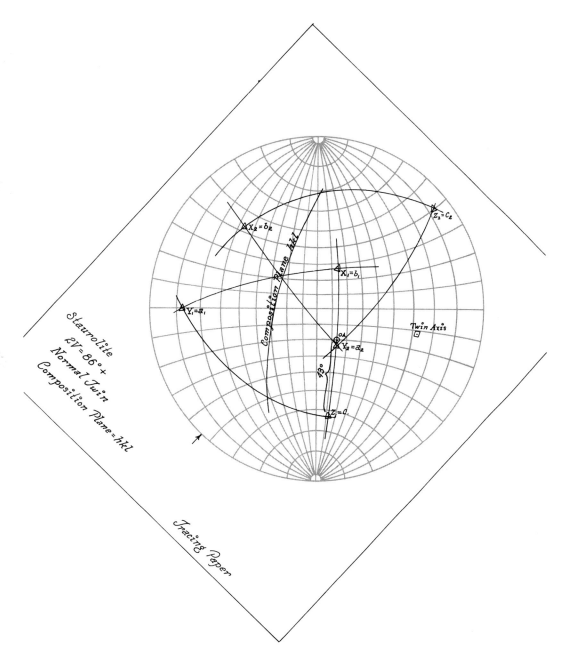

FIGURE 11-6. Staurolite twin. Staurolite is a pseudo-orthorhombic mineral with $a = Y$, $b = X$, and $c = Z$. By plotting X, Y, and Z for both halves of the twin and the twin composition plane, we see that the composition plane parallels no crystallographic direction and must, therefore, be $\{hkl\}$, probably $\{111\}$. When the composition plane is vertical N–S, both halves show equal birefringence on rotation about A_4, so the twin is normal. The twin axis is therefore normal to the composition plane, which is also the twinning plane.

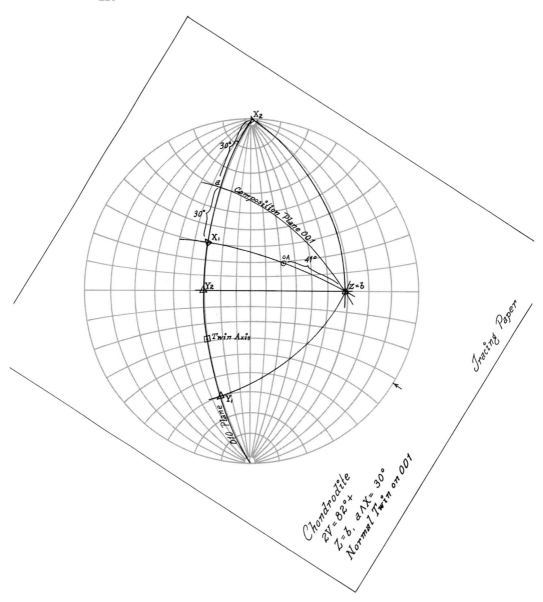

FIGURE 11-7. Chondrodite twin. Chondrodite is a monoclinic mineral that twins on {001}. Plotting X, Y, and Z for both halves of the twin and the {001} composition plane, we see that Z is the same for both halves ($Z = b$); the composition plane passes through b and through a in the {010} plane (i.e., the plane normal to b), so that $a = X = 30°$.

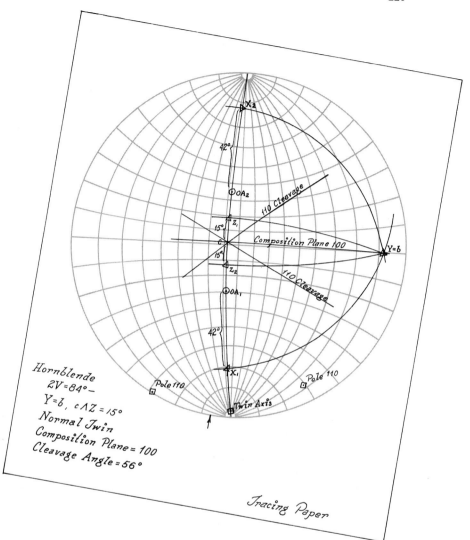

FIGURE 11-8. Hornblende twin. We know that hornblende is a monoclinic mineral with perfect prismatic {110} cleavages. By plotting X, Y, and Z for both sides of the twin and the traces of prismatic cleavages and composition plane, we can observe the following: (1) Y is the same for both twin halves, $Y = b$. (2) Cleavage traces intersect at c on the {010} plane (90° from b), and $c \wedge Z = 15°$. (3) Since the composition plane passes through (i.e., parallels) both b and c, it is the {100} plane. (4) By placing both cleavage poles on a great circle, we find the angle between cleavages to be 56°. (5) We observed the twin to be normal, when plotting the trace of the composition plane.

Twinning in Triclinic Minerals

Triclinic crystals form normal, parallel, and complex twins with composition planes usually parallel to a pinacoidal face, {100}, {010} or {001}. Such twins yield a stereographic pattern in which no optical directions coincide across the composition plane (Figs. 11-9 and 11-10). Many triclinic minerals, however, are almost monoclinic and yield patterns in which one optical direction lies very close to its counterpart from the opposite twin individual. Crystallographic directions are often difficult to locate but known cleavages or composition planes may be adequate for the location of a, b, or c; under favorable circumstances, angles between optical and crystallographic directions can be measured, and even crude measurements of axial angles (α, β, and γ) are possible.

Feldspar Studies

Early works by Fedorow and others show that they had an intense interest in applying the universal stage to the all-important task of distinguishing members of the complex feldspar group. Ionic substitutions in the feldspar framework cause changes in optical orientation, and numerous methods of feldspar identification are aimed at establishing some relationship between optical and crystallographic directions. The universal stage is used to establish optical directions (X, Y, and Z), and should any known crystallographic planes be visible to establish any crystallographic orientation, they can be used as a basis for identification.

For most varieties of feldspar, twin planes furnish the key to crystallographic orientation. Laws of feldspar twinning are each defined by the appropriate composition plane and twin axis, which may define a normal, parallel, or complex twin.

A useful relationship between optical and crystallographic directions can be established by any one of several simple methods.

Method I is based on the angular relationship of the {010} normal (crystallographic direction) to X, Y, and Z (optical directions).

1. A plagioclase grain showing albite twinning is selected, preferably one with a relatively large untwinned area. Albite twinning is recognized as the only repeated normal twin.

TABLE 11-1. Feldspar Twin Laws

Twin Name	Composition Plane*	Twin Axis	Type	Occurrence
Simple Twinning				
Monoclinic Feldspars				
Carlsbad	(010)	c axis	parallel	very common
Manebach	(001)	$\perp$(001)	normal	uncommon
Baveno	(021)	$\perp$(021)	normal	uncommon
Ala A	(001)	a axis	parallel	uncommon
Ala B	(010)	a axis	parallel	uncommon
Triclinic Feldspars				
Carlsbad	(010)	c axis	parallel	very common
Manebach	(001)	$\perp$(001)	normal	uncommon
Baveno	(021)	$\perp$(021)	normal	uncommon
Repeated Twinning				
Triclinic Feldspars				
albite	(010)	$\perp$(010)	normal	very common
Ala A	(001)	a axis	parallel	uncommon
Ala B	(010)	a axis	parallel	uncommon
pericline	Zone of b axis†	b axis	parallel	common
acline	(001)	b axis	parallel	common
albite-Carlsbad	(010)	$\perp c$ axis in (010)	complex	common
albite-Ala B	(010)	$\perp a$ axis in (010)	complex	uncommon
Manebach-Ala A‡	(001)	$\perp a$ axis in (001)	complex	uncommon
Manebach-acline‡	(001)	$\perp b$ axis in (001)	complex	uncommon

*A set of twin types exists with (010) composition plane and an essentially equivalent set with (001) composition plane. By far the most common twins are formed on (010), but twins on (001) are fairly common in calcic feldspars. A similar set of twin laws is theoretically possible on (100); such sets have been reported but appear exceedingly rare.

†The composition plane of pericline twinning is a ($h0l$) section, called the rhombic section (i.e., parallel to the b-axis), which varies in orientation with composition (Fig. 11-11). For an intermediate andesine composition, pericline and acline twinning are identical.

‡Since $\angle\gamma$ ranges from 87° for albite to $91\frac{1}{2}$° for anorthite, b and $\perp a$ are essentially coincident, never differing by more than 3°. Acline and Manebach-Ala A are practically indistinguishable. Similarly, Ala A and Manebach-acline are indistinguishable.

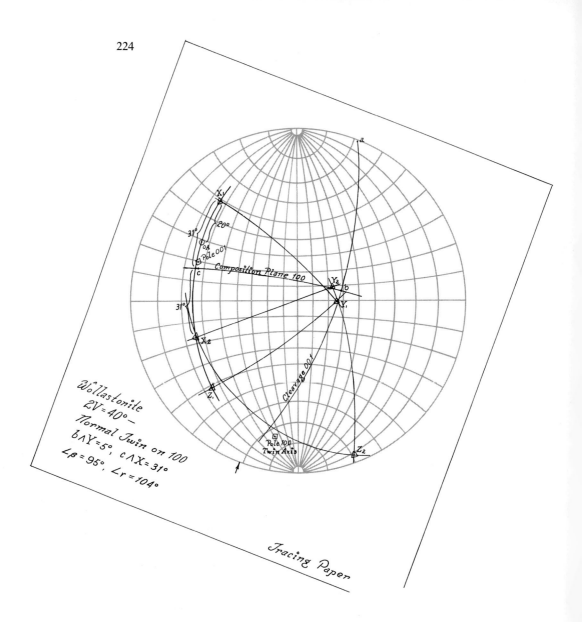

FIGURE 11-9. Wollastonite twin. Wollastonite is triclinic with perfect cleavage parallel to {100} and less perfect parallel to {001} and {102}. It is known to twin on {100}. A plot of X, Y, and Z for both crystal halves shows no coincidence of optical directions, although Y_1 and Y_2 are close together. The existing twin is normal, and its composition plane parallels an excellent cleavage, presumably {100}. A second obvious cleavage, presumably {001}, intersects the {100} plane at b, which lies only 5° from each Y ($b \wedge Y = 5°$). Both b and c must lie on the {100} plane, and since $\angle \alpha = 90°$, c is located only 5° from the normal to {001} and $\angle \beta = 95°$. Knowing $\angle \alpha = 90°$ (given), $\angle \beta = 95°$, and the trace of {001}, we can find the approximate point on {001} at which c is 95° from {001}, and we find a about 76° from b and $\angle \gamma = 103°$; c is half way between X_1 and X_2, and $c \wedge X = 31°$.

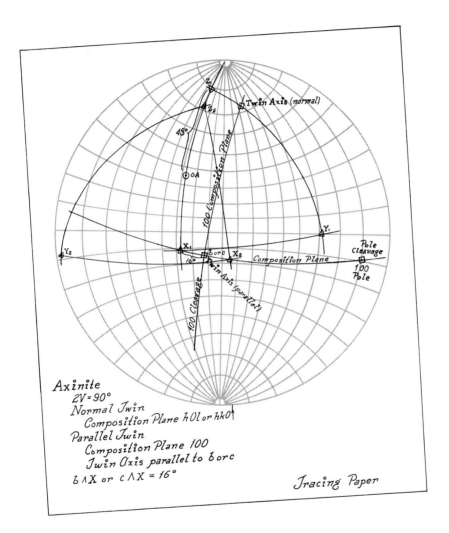

FIGURE 11-10. Axinite twins. Axinite is a triclinic mineral with distinct cleavage on {100}. A single crystal is observed to possess both normal and parallel twinning. The composition plane for parallel twinning parallels the distinct cleavage {100}, and the twin axis is located at the intersection of {100} with the composition plane of the normal twin. Since the twin axis of the parallel twin must parallel some crystallographic axis in the {100} plane, this intersection is either b or c, and the composition plane of the normal twin also parallels this crystallographic axis with Miller index {$h0l$} or {$hk0$}.

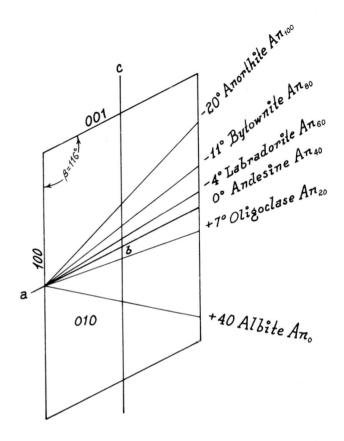

FIGURE 11-11. Composition plane for pericline twinning—Variation with plagioclase composition. The composition plane for pericline twinning parallels the *b*-crystallographic axis, ranging from 20° above the basal plane (001) for anorthite to 40° below (001) for albite. At a composition of An_{40} (andesine), pericline and acline twinning are synonymous.

2. Optical directions X, Y, and Z are plotted for one set of albite twin laminae.

3. The pole of the composition plane {010} (i.e., the twin axis) can be plotted with good accuracy, since albite twinning is normal and twin laminae show uniform birefringence when {010} is vertical.

4. The entire stereogram is rotated so that Y is vertical and the XZ-plane

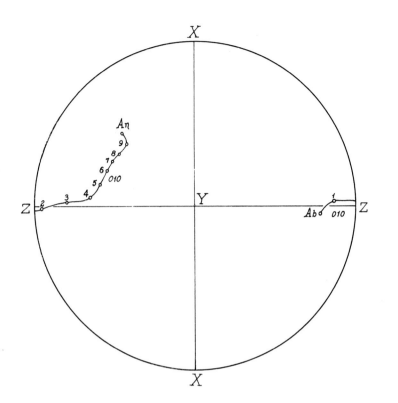

FIGURE 11-12. Migration curve for the (010) pole in plagioclase. For a given plagioclase composition, the spacial relationship between optical directions and crystallographic directions is fixed. This relationship is expressed by a stereogram showing X, Y, and Z and the pole of (010), the twinning axis of an albite twin. If stereogram and migration curve are the same diameter and the stereogram is rotated so that Y is at the center, proper orientation of the stereogram causes the (010) pole to fall on, or near, the migration curve that indicates plagioclase composition.

horizontal (Fig. 9-12). The stereogram now shows the position of the {010} normal for a standard orientation of X, Y, and Z.

5. Figure 11-12 shows migration of the {010} normal with plagioclase composition. The figure should be enlarged to the same diameter as the stereogram so the tracing can be superimposed on the figure to show where the {010} normal lies on the curve to betray plagioclase composition (i.e. anorthite content). The stereogram may be superimposed in four positions with X and Z properly

aligned on the periphery of the figure; two positions require the tracing to be turned over, face down. The {010} pole may fall near the migration curve in two of these positions, one indicating high anorthite content and the other low. Above 20 percent anorthite, the refraction index of plagioclase is greater than 1.54 (the index of balsam), below 20 percent, it is less than 1.54.

Keep in mind that a migration curve is not really a mere line but a band of finite thickness, widened by chemical impurities and other unknown factors. An accuracy of ± 5 percent anorthite can be expected for compositions more calcic than 40 percent, and an accuracy of ± 10 percent anorthite can be expected for more sodic compositions.

Method II (F. J. Turner, 1947) also makes use of the angular relationship between the normal to a composition plane (crystallographic direction) and optical directions X, Y, and Z.

1. A plagioclase grain showing any type of twin may be used (avoid rare Baveno twins, in which the composition plane lies nearly 45° to all other twin planes and cleavages).

2. Optical directions are plotted for both individuals of a twin (X_1, X_2, Y_1, Y_2, Z_1, and Z_2).

3. The normal (i.e., pole) to the twin composition plane ($\perp CP$) is plotted on the same stereogram.

4. At some rotation of the stereogram, X_1, X_2, and $\perp CP$ should lie on a great circle with the pole bisecting the angle between X_1 and X_2 (i.e., $X_1 \wedge CP = X_2 \wedge CP = X \wedge CP$). Measure and record $X \wedge CP$, $Y \wedge CP$, and $Z \wedge CP$.

5. Figure 11-13 shows the variation of $Z \wedge CP$ with plagioclase composition (percent An) for twins on both {010} and {001}. Any value of $Z \wedge CP$ identifies the composition plane of the twin as {010} or {001}, except near $Z \wedge CP = 43°$, and yields the plagioclase composition. Below about 30 percent anorthite, two compositions are shown by the graph, however, it may be remembered that compositions less calcic than 20 percent anorthite have refraction indices below 1.54 (balsam).

6. Figure 11-14 shows the variation of $X \wedge CP$ and $Y \wedge CP$ with composition for twins on (010) and (001). One curve may be used to confirm the composition derived from Figure 11-13.

This method of F. J. Turner (1947), revised by D. B. Slemmons (1962), turns out to be rather complex when all possible ambiguities are considered; the student is therefore encouraged to consult the original works.

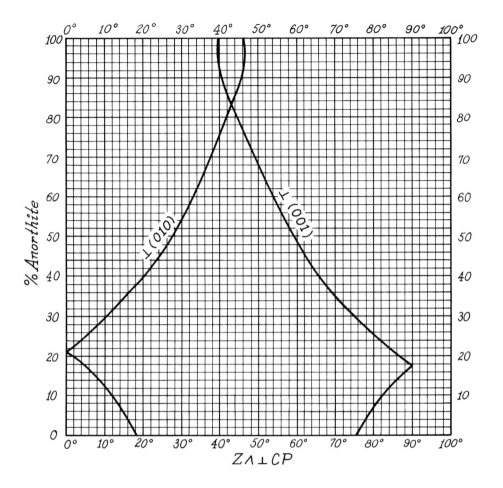

FIGURE 11-13. Variation of $Z \wedge \perp CP$ with plagioclase composition. For one individual of a plagioclase twin, the angle between optical direction Z_1 and the normal to the composition plane (i.e., $Z_1 \wedge \perp CP$) should equal the corresponding angle for the other twin individual (i.e., $Z_1 \wedge \perp CP = Z_2 \wedge \perp CP = Z \wedge \perp CP$). This angle ($Z \wedge \perp CP$) varies with plagioclase composition, as shown by separate curves for {010} and {001} composition planes. As essentially all plagioclase twins are united on either {010} or {001}, almost any twin may be used for measurement. The figure gives the composition plane of the measured twin as well as anorthite content. These curves should be used in combination with Figure 11-14. The composition planes of Baveno and pericline twins (Fig. 11-11) are neither {010} nor {001}, but these twins are usually obvious because their composition planes normally make highly oblique angles to cleavages and other twin planes.

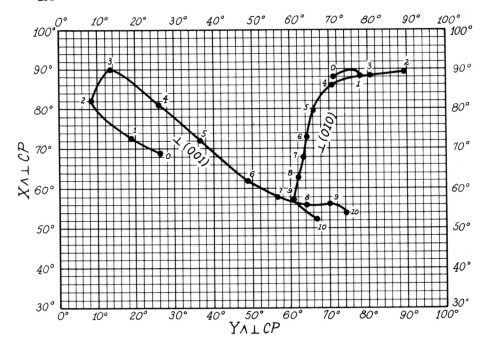

FIGURE 11-14. Variation of $X \wedge \perp CP$ and $Y \wedge \perp CP$ with plagioclase composition. For one individual of a plagioclase twin, the angle between optical direction X_1 and the normal to the composition plane (i.e., $X_1 \wedge \perp CP$) should equal the corresponding angle for the other twin individual (i.e., $X_1 \wedge \perp CP = X_2 \wedge \perp CP = X \wedge \perp CP$) and likewise for the Y-optical direction (i.e., $Y_1 \wedge \perp CP = Y_2 \wedge \perp CP = Y \wedge \perp CP$). These angles ($X \wedge \perp CP$ and $Y \wedge \perp CP$) vary with plagioclase composition, as is shown by separate curves for the $\{010\}$ and $\{001\}$ composition planes. The small numbers on the migration curves indicate the percentage of anorthite composition, from zero percent at 0 to 100 percent at 10.

CHAPTER 12

Preparation of the Sample

Mineral Fragments

Purpose and Advantages of Studying Fragments

For systematic identification of individual minerals, crushed fragments are superior to thin sections. With a single immersion mount of mineral fragments, the following measurements and observations may be made.

With uncrossed nicols:

   1. Color and pleochroism, if present.

   2. Cleavage and fragment shape or form.

   3. Relief, for comparison of refraction index with that of immersion medium.

With crossed nicols:

   1. Maximum birefringence.

   2. Extinction angle, if appropriate.

   3. Sign of elongation, if appropriate.

   4. Anomalous interference colors, if present.

   5. Twinning, if present.

With crossed nicols and conoscopic illumination:

1. Uniaxial or biaxial nature of crystal
2. Optic sign.
3. Magnitude of 2V, if appropriate.
4. Dispersion (optic axis and bisectrix), if present.

By placing mounted fragments in appropriate immersion liquids in succession, it is possible to measure the numerical value of the principal indices of refraction (n or n_ϵ and n_ω or n_α, n_β, and n_γ); therein lies the main advantage in studying fragments rather than sections.

Tables of optical data are based upon (1) isotropic, uniaxial, and biaxial groups, (2) optical sign, and (3) principal refractive indices.

The Preparation of Fragment Mounts

Mineral fragments for microscopic study are best prepared by crushing rather than grinding so that cleavages and structure habits may be as obvious as possible. Fragments about 0.1 mm in size are the most convenient to work with.

Mineral fragments are usually studied in temporary immersion-oil mounts, but permanent mounts may be prepared by melting a spot of cooked Canada balsam or Lakeside 70 cement on a glass slide, sprinkling a few mineral fragments into the liquid resin, and covering them with a cover glass. Such mounts show the general appearance of specific mineral varieties and may be examined by standard universal stage procedures.

Rock Thin Sections

Purpose and Advantages of Studying Thin Sections

A rock thin section shows mineral content, abundance and association, grain size, alteration, rock structures and textures, and is a permanent record of a given rock that may be filed for future reference.

The Preparation of Rock Thin Sections

A rock thin section is, indeed, a very thin section of a rock, and its preparation requires some inherent skill combined with a little experience. Technology may supply better cut-off saws, laps, abrasives, and cementing agents, but the quality of the final thin section is a function of the skill of the man who makes it. The past decade has seen the introduction of thin-sectioning machines that claim to cut dozens of sections per hour, but the user must prepare the section to be cut by a few basic steps.

1. Thin sections are made on petrographic microscope slides (26 mm × 46 mm about 1.25 mm thick)* and covered by thin, No. 1, cover glasses (0.13 mm to 0.17 mm). One side of the slide may be frosted by grinding on a horizontal lap with No. 600 silicon carbide abrasive. The roughened surface makes a better bond with the cementing agent and reduces the chance that the section will be torn from the slide during the final grinding stages.

2. A rock section about 20 mm × 25 mm and about 3 mm thick and of desired orientation is cut from the rock sample. Cutting is normally accomplished with a small trim saw equipped with a diamond-studded blade that is cooled by a liquid lubricant.† Soft rocks may be held against the blade by hand, but very hard rocks like chert or dense quartzite should be held in a mechanical vise, because cutting is slow and difficult. A slight twist of the rock will crimp a saw blade, rendering it useless.

3. If the rock section is poorly consolidated, highly fractured, or porous, it should be artificially consolidated and the void spaces filled before proceeding further. Adequate consolidation can often be accomplished by spreading raw balsam over the rock surface and cooking the section on a hot plate (150°C) until the cold balsam is no longer sticky. Very soft and friable materials may require more elaborate procedures (see References, p. 244).

4. For grinding the section to its proper thickness, a pair of horizontal cast iron laps should be available. One lap is used with No. 100 and No. 220 silicon carbide abrasive for initial coarse grinding; the other is reserved for final grinding with No. 600 abrasive only. In some laboratories a very slowly rotating lap is used for final grinding and in others the final grinding is done by hand on a glass plate. Bonded diamond wheels, used as laps or surface grinders, may speed up the initial grinding stages.

5. The prepared side of the rock section (i.e., the side that is to be bonded to the slide) is ground as smooth and as flat as possible with No. 600 abrasive.

---

*For oil immersion objectives, a slide thickness of 1.0 mm is preferable, and, for certain "automatic" grinding techniques, slide thickness may be critical.

†Water, the most common coolant, allows greater blade wear than a mixture of soluble oil (e.g., "Caloil") and water; the mixture should be sufficiently rich to feel slippery. Kerosene or kerosene and motor oil are good coolants but are flammable.

This step is important, for the surface must not be even slightly convex. A single finger placed in the center of the section during its final grinding will usually prevent rounding.

6. The rock section must be cemented to the glass slide with great care to avoid difficulties in the final stage of grinding. Just before cementing, the rock section and the slide are placed on a hot plate (300°C) with the prepared surface of the section and the frosted side of the slide facing upward. Cementing resin* is then melted onto both slide and section.

The glass slide must be lifted from the hot plate with forceps as soon as the resin has melted; it is then inverted and placed on the rock section, and both are moved from the hot plate onto a smooth, hard surface. Assuming the rock section is not porous, or the pore spaces have been previously filled (see step 3), the slide may be pressed down firmly and moved about on the section until all bubbles disappear from between the two surfaces. A small weight is placed on the couple until the resin has cooled. If a film of air or an air bubble is trapped between slide and section, there will most likely be trouble in the final grinding stages, since there is no support for the rock section at the bubble. A break at this point will release large mineral grains onto the lap surface, and these will tear out additional grains.

7. Coarse grinding or sawing is used to reduce the section to about 0.1 mm. The cemented section can usually be returned to the diamond saw and, with the aid of a parallel guide, be reduced to 0.5 mm or less. Coarse grinding with No. 100 and No. 220 abrasive can further reduce the section to about 0.1 mm. It is in this step that the widely publicized "automatic" thin-sectioning machines prove their worth by sawing or grinding the section to nearly 0.03 mm without manual grinding. The time saved on a single section is almost negligible but is significant for mass-production procedures.

8. Final grinding to 0.03 mm with No. 600 abrasive is the critical step, and it is here that the beginner may have to begin anew. The success of this step, however, depends in part on the care taken in the steps that precede it.

---

*Lakeside 70 is a very widely used cementing resin which comes in solid sticks which are melted directly onto slide and rock section. It is available from Ward's Natural Science Establishment.

Cooked Canada balsam is the original cementing resin. It is prepared by heating a shallow tin dish of raw Canada balsam on a hot plate at 300°C. When a cooled sample retrieved on a metal point is brittle but still plastic enough to yield slightly to pressure applied with a knife edge or fingernail before fracturing, the balsam is cooled and cracked out of the dish as large pieces.

Numerous synthetic resins have been tried and recommended, and the student may wish to investigate one or more (see References, p. 243).

Final grinding may be completed on a revolving lap or on a glass plate, depending on skill and experience. All grinding is done wet. Coarse abrasives are usually sprinkled onto a wet lap from a shaker, but the No. 600 abrasive should be mixed with water in a flask with stopper and penetrating tube and squirted onto the wet lap like hair tonic. Friction between finger and slide must always exceed friction between lap and section, otherwise the slide will be pulled away. This is controlled by application of the water-abrasive mixture from the flask. *Caution:* All grinding should be done with the side of the thumb or the middle finger held directly in the center of the section.* The pressure applied determines the rate of grinding. Final grinding requires very light pressure. With a little practice, the section can be made to ride on a film of water over the revolving lap with essentially no grinding at all. Each time the lap is charged with fresh abrasive from the flask, grinding is rapid. With a few revolutions of the lap, however, the tiny abrasive grains become rounded and reduced in size, and grinding becomes slower and more easily controlled. The last few seconds of grinding are therefore best done without fresh abrasive, and sometimes it is wise to add a few drops of water to prevent the plate from drying.

The section should be examined after every few turns of the lap to avoid wedging or to avoid grinding away some portion of the section. At thicknesses less than 0.1 mm the section will be quite transparent; a quick glance at the section with back light from a lamp or window will show which area is thickest. A few turns of the lap or a few revolutions on the glass plate with a single finger held directly over the thicker area will usually restore the section to uniform thickness. With considerable practice, one can tell when a section is essentially 0.03 mm thick without optical examination. The student, however, will find it advisable to evaluate his progress by examining the section periodically between crossed nicols during the final stages of grinding. It is very unlikely that he will grind the section too thin, but he may well grind it completely away if he is asleep at the lap. The thickness of a section is measured by the retardation colors of recognizable minerals. Most rocks contain either quartz or feldspar, both of which have inherent maximum birefringence of 0.007 to about 0.011. At a thickness of 0.03 mm, no quartz grain can show a retardation that exceeds 270 mμ (pale first-order yellow), and feldspar grains show first-order grey, white, or yellow.

---

*Should the beginner hold the slide by its edges with his thumb and middle finger resting against the abrasive surface, he will soon discover that he can grind away his skin without feeling any pain—until the wound dries!

Since clay minerals often expand and contract with wetting and drying, highly altered rocks may be ground by using kerosene in place of water.

9. Excess cementing resin is scraped from around the rock section, and the section is wiped with a soft tissue and acetone to remove abrasive grains imbedded in the resin.

10. To mount the cover glass, fragments of cooked Canada balsam are dissolved in xylene to form a thick syrupy mixture. The section is placed on a warming plate (60°C) and the cover glass on a hot plate (150°C); both are then spread with the balsam-xylene mixture. The cover glass is cooked for one to three minutes and inverted onto the section and pressed down gently to remove air bubbles. If the mixture is sufficiently viscous, the section may be almost immediately wiped with xylene or acetone to remove excess balsam.

11. The uncovered end of the glass slide can now be labeled by writing directly on the frosted surface.

12. The thin section is made ready for use by washing it with water and a cleanser and drying it. Care must be exercised in using a new thin section between hemispheres of a universal stage, where the cover glass tends to slide over the section.

References

Textbooks—Optical Crystallography

Bloss, F. D. (1961). An Introduction to the methods of optical crystallography: Holt, Rinehart and Winston, New York.

El-Hinnawi, E. (1966). Methods in chemical and mineral microscopy: Elsevier, New York.

Gay, P. (1967). An introduction to crystal optics: Longmans, Green and Co., Ltd., London.

Hartshorne, N. H., and A. Stuart (1970). Crystals and the polarizing microscope: American Elsevier, New York.

—— (1970). Practical optical crystallography, 4th Edition: American Elsevier, New York.

Kerr, P. F. (1959). Optical mineralogy, 3rd Edition: McGraw-Hill, New York.

Marshall, C. E. (1949). Introduction to crystal optics: Cook, Troughton and Simms, Yorkshire, England

Shubnikov, A. V. (1960). Principles of optical crystallography: Consultants Bureau, New York.

Wahlstrom, E. E. (1969). Optical crystallography, 4th Edition: Wiley, New York.

Winchell, A. N. (1949). Elements of optical mineralogy, Part 1-Principles and methods, 5th Edition: Wiley, New York.

Handbooks of Petrographic Microscopy

Burri, C. (1950). Das Polarisationsmikroskop: Birkhaüser, Basel.

Freund, H. (1957). Handbuch der Microskopie in der Technik, 8 Bände: Umschau Verlag, Frankfurt am Main.

Hallimond, A. F. (1953). Manual of the polarizing microscope: Cooke, Thoughton and Simms, York, England.

Hecht, F. und M. H. Zacher (1954). Handbuch der mikrochemischen Methoden, Band 5: Springer, Wien.

Johannsen, A. (1918). Manual of petrographic methods: McGraw-Hill, New York.

Pockels, F. (1906). Lehrbuch der Kristalloptik: B. G. Teubner, Berlin.

The Microscope

Francom, M. (1964). Progress in microscopy: Pergamon Press, Oxford.

Hallimond, A. F. (1953). Manual of the polarizing microscope: Cooke, Thoughton and Simms, York, England.

Martin, L. C. (1966). The theory of the microscope: Blackie and Son, Ltd., Glasgow.

Shurcliff, W. A. (1962). Polarized light: Harvard Univ. Press, Cambridge, Mass.

Refractometry

Methods and Procedures

Allen, R. M. (1954). Practical refractometry by means of the microscope: R. P. Cargille Lab. Inc., New York.

Becke, F. (1893). Sitzber. Akad. Wiss. Wien, C11, Abt. 1, p. 358.

Emmons, R. C. (1929). The double variation method: Am. Mineralogist, *14*: 482.

────── and R. M. Gates (1948). Use of Becke line colours in refractive index determination: Am. Mineralogist, *33*: 612.

Fairbairn, H. W. (1943). Gelatin coated slides for refractive index immersion mounts: Am. Mineralogist, *28*: 396.

Harrison, R. K., and G. Day (1963). A continuous monochromatic interference filter: Mineralogical Mag., *33*: 517.

Olcott, G. W. (1960). Preparation and use of a gelatin mounting medium for repeated oil immersion of minerals: Am. Mineralogist, *45*: 1099.

Saylor, C. P. (1935). Accuracy of microscopical methods for determining refractive index by immersion. J. Res. Natl. Bur. St. C.: *15*: 277.

Immersion Liquids

Barth, T. (1929). Some new immersion melts for high refraction: Am. Mineralogist, *14*: 358.

Buerger, M. J. (1933). The optical properties of ideal solution immersion liquids: Am. Mineralogist, *18*: 325.

Butler, R. D. (1933). Immersion liquids of intermediate refraction (1.450-1.630): Am. Mineralogist, *18*: 386.

Darneal, R. L. (1948). Immersion media containing methylene iodide: Am. Mineralogist, *33*: 346.

Fujii, T., and F. D. Bloss (1962). Some properties of α-monochloro-naphthalene-diiodomethane immersion media: Am. Mineralogist, *47*: 267.

Kaiser, E. P., and W. Parrish (1939). Preparation of immersion liquids: Ind. Eng. Chem. (Anal. Edition), *11*: 560.

Meyrowitz, R. (1952). A new series of immersion liquids: Am. Mineralogist, *37*: 853.

———— (1955). A compilation and classification of immersion media of high index of refraction: Am. Mineralogist, *40*: 398.

———— (1956). Solvents and solutes for the preparation of immersion liquids of high index of refraction: Am. Mineralogist, *41*: 49.

———— and F. S. Larsen, Jr. (1951). Immersion liquids of high refractive index: Am. Mineralogist, *36*: 746.

Weaver, C. F., and T. N. McVay (1960). Immersion oils with indices of refraction from 1.292 to 1.411: Am. Mineralogist, *45*: 469.

Wilcox, R. E. (1964). Immersion liquids of relatively strong dispersion in the low refractive index range (1.46-1.52): Am. Mineralogist, *49*: 683.

West, C. D. (1936). Immersion liquids of high refractive index: Am. Mineralogist, *21*: 245.

Refractometers

Coats, R. R. (1960). Method of minimizing damage to refractometers from use of arsenic tribromide liquids: Am. Mineralogist, *45*: 903.

Jelley, E. E. (1934). A microrefractometer and its application in microchemistry: J. Roy. Microscop. Soc. (London), *54*: 234.

Optical Crystallography

The Indicatrix

Joel, N., and I. Garaycochea (1957). The "extinction curve" in the investigation of the optical indicatrix: Acta. Cryst., *10*: 399.

Fletcher, L. (1892). The optical indicatrix and the transmission of light in crystals: H. Froude, London.

Phemister, T. C. (1954). Fletcher's indicatrix and the electromagnetic theory of light: Am. Mineralogist, *39*: 172.

Axial Angle and Refraction Indices

Gravenor, C. P. (1951). A graphical simplification of the relationship between $2V$ and N_x, N_y and N_z: Am. Mineralogist, *36*: 162.

Lane, J. H., Jr., and H. T. U. Smith (1938). Graphic method of determining the optic sign and true axial angle from refractive indices of biaxial minerals: Am. Mineralogist, *23*: 457.

McAndrew, J. (1963). Relation of optic axial angle with the three principal refractive indices: Am. Mineralogist, *48*: 1275.

Mertie, J. B., Jr. (1942). Nonograms of optic angle formulae: Am. Mineralogist, *27*: 538.

Smith, H. T. U. (1937). Simplified graphic method of determining approximate axial angle from refractive indices of biaxial minerals: Am. Mineralogist, *22*: 675.

Swift, P. Mel. (1954). The indirect determination of beta index of refraction and $2V$: Am. Mineralogist, *39*: 838.

Tobi, A. C. (1956). A chart for the measurement of optic axial angles: Am. Mineralogist, *41*: 516.

Wright, F. E. (1951). Computation of the optic axial angle from the three principal refractive indices: Am. Mineralogist, *36*: 543.

Birefringence

Pittmann, A. (1951). On the determination of birefringence in the case of high retardation: Bull. Fac. Sci., Alexandria Univ., *1*: 1.

Interference Figures

Becke, F. (1905). Die Skiodromen: Tsch. Min. Pet. Mitt., *24*: 1.

Grabar, D. C. (1962-63). Measurement of optical axial angles on Bxa interference figures: Microscope, *13*: 288.

Kamb, W. B. (1958). Isogyres in interference figures. Am. Mineralogist, *43*: 1029.

Winchell H. (1946). A chart for measurement of interference figures: Am. Mineralogist, *31*: 43.

Wright, F. E. (1923). The formation of interference figures: J. Opt. Soc. Am., 7: 805.

Dispersion

Bryant, W. M. D. (1941). Optical crystallographical studies with the polarizing microscope III, measurement of several types of selective dispersion in organic compounds: J. Am. Chem. Soc., *63*: 511.

——— (1943). Axial dispersion with change in sign: J. Am. Chem. Soc., *65*: 96.

Color and Pleochroism

Ehlers, J. (1898). Die Absorption des Lichtes in einigen pleochroitischen Kyrstallen: Neues Jahrb., *11*: 259.

Mandarino, J. A. (1959). Absorption and pleochroism: two much neglected optical properties of crystals: Am. Mineralogist, *44*: 65.

Crystal Structure

Batsanov, S. S. (1961). Refractometry and chemical structure: Consultants Bureau, New York.

Boky, G. B., and S. S. Batsonov (1955). Crystal-optical method of determining the structure of complex compounds: Bull. Acad. Sci (USSR), Div. Chem. Sci., 173.

Universal Stage

Methods and Procedures

Adams, W. G. (1875). A new polariscope: Phil. Mag., *50*: 13.

Berek, M. (1923). Neue Wege zur Universalmethode: Neues Jahrb., Mineral., *48*: 34.

Emmons, R. C. (1943). The universal stage: Geol. Soc. Am., Mem. 8.

von Fedorov, E. (1896). Universalmethode und Feldspathstudien: Zeits. Krist. Min., *26*: 225.

Joel, N., and I. D. Muir (1958). New tecyniques for the universal stage I and II: Mineralogical Mag., *31*: 860.

Kamb, W. B. (1962). Refraction correction for universal stage measurements, 1. Uniaxial crystals: Am. Mineralogist, *47*: 227.

Kleeman, A. W. (1952). Nomograms for correcting angle of tilt of the universal stage: Am. Mineralogist, *37*: 115.

Nikitin, W. W. (1936). Die Federow-Methode: Borntraeger, Berlin.

Reinhard, M. (1931). Universal Drehtischmethoden: Verlag Von B. Wepf und Cie., Basel.

Sarantschina, G. M. (1963). Die Federow-Methode: Verlag der Wissenschaften, Berlin.

Turner, F. J., and C. Gilbert (1949). Use of the U-stage in sedimentary petrography: Am. J. Sci., *247*: 1.

Wenban-Smith, A. K. (1967). A computer program for determining optical crystallographic directions directly from universal stage readings: Canad. Min., *9*: 269.

Wülfing, E. A. (1921-1924). Untersuchungsmethoden, Bd. I. *In* H. Rosenbusch, Mikroskopische Physiographie der petrographisch wichtigen Mineralien.

Axial Angle

Dodge, T. A. (1934). The determination of optic angle with the universal stage: Am. Mineralogist, *19*: 62.

Fairbairn, H. W., and T. Podolsky (1951). Notes on precision and accuracy of optic angle determination with the U-stage: Am. Mineralogist, *36*: 823.

Joel, N., and I. D. Muir (1964). Extinction measurements for the determination of $2V$ with the universal stage: Am. Mineralogist, *49*: 286.

Munro, M. (1963). Errors in the measurement of $2V$ with the universal stage: Am. Mineralogist, *48*: 308.

────── (1966). The measurement of large optic angles with the universal stage: Mineralogical Mag., *35*: 763.

Tocher, F. E. (1964). A new universal stage extinction method: The determination of $2V$ when neither optic axis is directly observable: Mineralogical Mag., *33*: 1038.

Conoscopic Methods

Hallimond, A. F. (1950). Universal stage methods: Mining Mag., *83*: 12 and 77.

Schumann, H. (1941). Über den Anwendungsbereich der konoskopischen Methodik: Forts. Mineralogy, *25*: 217.

────── (1951). Orthoskopische und konoskopische Boebachtungsweise im Universaldrehtish. Mikroscopie, *6*: 104.

Application of U-stage to Mineral Groups

Stemmons, D. B. (1962). Determination of volcanic and plutonic plagioclase using a three- or four-axis universal stage: G. S. A. Spec. Paper No. 69.

Turner, F. J. (1942). Determination of extinction angles in monoclinic pyroxenes and amphiboles: Am. J. Sci., *240*: 571.

——— (1947). Determination of plagioclase with the four-axis U-stage: Am. Mineralogist, *32*: 389.

Petrofabrics

Donn, W. L., and J. A. Shimer (1958). Graphic methods in structural geology: Appleton-Century-Crofts, Inc., New York.

Fairbairn, H. W. (1949). Structural petrology of deformed rocks: Addison-Wesley, Reading, Mass.

Knopf, E. B., and E. Ingerson (1938). Structural petrology: G. S. A. Memoir 6.

Turner, F. J., and L. E. Weiss (1963). Structural analysis of metamorphic tectonics: McGraw-Hill, New York.

Preparation of the Sample

General

Bishop, A. C. (1962). An improved technique for producing thin rock-sections: Mineralogical Mag., *33*: 274.

Meyer, C. (1946). Notes on the cutting and polishing of thin sections: Econ. Geol., *41*: 166.

Milner, H. B. (1962). Sedimentary petrography. Volume I—Methods of sedimentary petrography: Macmillan, New York.

Reed, F. S., and T. D. Mergener (1953). Preparation of rock thin sections: Am. Mineralogist, *38*: 1183.

Rowland, E. O. (1953). A rapid method for the preparation of rock thin sections: Mineralogical Mag., *30*: 255.

Cementing Resins

Fischer, G. (1955). DMS-balsam ein neues Einbettungsmittle für optische Präparate: Mikroskopie, *10*: 333.

Von Huene, R. (1949). Notes on Lakeside No. 70 transparent cement: Am. Mineralogist, *34*: 125.

Wright, H. G. (1964). The use of epoxy resins in the preparation of petrographic thin sections: Mineralogical Mag., *33*: 931.

Special Techniques

Amstutz, G. C. (1960). The preparation and use of polished thin sections: Am. Mineralogist, *45*: 1114.

Bauman, H. N., Jr. (1957). Preparation of petrographic sections with diamond wheels: Am. Mineralogist, *42*: 416.

Bennett, R. L. (1958). Evaporite sections: J. Inst. of Sci. Tech., *4*: 358.

Taylor, J. C. M. (1960). Impregnation of rocks for sectioning: Geol. Mag., *97*: 261.

Tables of Optical Constants and Mineral Descriptions

Deer, W. A., R. A. Howie, and J. Zussman (1962). Rock forming minerals. Volume I—Ortho- and ring silicates. Volume II—Chain silicates. Volume III—Sheel silicates. Volume IV—Framework silicates. Volume V—Nonsilicates: Longmans, London.

Heinrich, E. W. (1965). Microscopic identification of minerals: McGraw-Hill, New York.

Larsen, E. S., and H. Berman (1934). The Microscopic determination of the nonopaque minerals: U. S. Geol. Survey Bull. 848.

Tröger, W. E. (1952). Tabellen zur optischen Bestimmung der gesteinsbeldenden. Minerale: E. Schweizebart'sche Verlags., Stuttgart.

——— (1959). Optische Bestimmung der gesteinsbeldenden Minerale. Teil I—Bestimmungstabellen: E. Schweizebart'sche Verlags., Stuttgart.

Winchell, A. N. (1951). Elements of optical mineralogy. Part II—Descriptions of minerals: Wiley, New York.

——— (1954). The optical properties of organic compounds: Academic Press, New York.

Winchell, N. W., and A. N. Winchell (1964). The microscopical characters of artificial inorganic solid substances. Optical properties of artificial minerals: Academic Press, New York.

Index

Aberration
  chromatic, 29–31
  spherical, 20, 29–31
Abrasives, 233–235
Airy disc, 32
Amphiboles
  orientation angle, 213
  twinning of, 217
Analyzer (upper nicol), 18, 26–27
  alignment of, 44–46
  polarization directions of, 26
Angle
  (ρ), 183
  (2E), 137
  see also Critical angle, Optical angle, Orientation angle
Angstrom unit, 7
Anisotropism, 7
  of biaxial crystals, 121
  test for, 72
  of uniaxial crystals, 75–79
Arm and base, microscope, 35
Azimuth angle, 183

Becke, F., 50
Becke line, 50–54
Bertrand lens, 23, 105
Biotite, absorption of, 44–45

Birefringence, 78, 98, 106
  of biaxial minerals, 135–136
  cause of, 101–102, 135–136
  chart of (follows 100)
  definition, 92
  positive and negative, 101n
  quantitative measurement of, 100–101
  of uniaxial minerals, 101–102
Bisectrix
  acute and obtuse, 125, 145–149
Bisectrix dispersion, 154–157
Bohr, Niels, 2
Brewster, Sir David, 14
Brewster's Law, 15

Calcite
  double refraction of, 77–79
  structure of, 76–77
  see also Polarizing prisms
Canada balsam, 232, 236
  cooking, 134n
  refractive index, 48
Carbonate minerals, identification of, 201–203
Cleavage
  of biaxial crystals, 163–168, 215
  of isometric crystals, 72
  of uniaxial crystals, 119–120, 196–197

Color, 4, 9
  of biaxial minerals, 159
  of isotropic minerals, 72
  of uniaxial minerals, 115–116
Complementary colors, 9
Compensators, Berek and slot, 101
Condensing lens, 37–38, 103
  for universal stage, 177
Conoscopic illumination, 103–105, 177, 196, 207–209
Cover glasses
  mounting of, 236
  thickness of, 32, 233
Critical angle, 12
Crystal structure
  of calcite, 76–77
  as a cause of anisotropism, 75–77
  as a cause of birefringence, 101–102, 135–136
  of isometric crystals, 70

Depth of focus, 34
Diamond saw, 233
Dichroism, 116
Dispersion of light, 7
  in biaxial interference figures, 154–157, 209
  bisectrix, 154–157
  bisectrix crossed, 157
  bisectrix horizontal, 156–157
  bisectrix inclined, 156
  fringes, 49–50, 54–56
  high-dispersion method, 53–57
  optic axis, 154
  in triclinic interference figures, 157
  see also Immersion liquids
Double refraction, 78
  see also Calcite
Duc de Chaulnes, law of, 78

Einstein, Albert, 2
Electromagnetic spectrum, 7, 8
  see also Waves
Elongation, sign of
  biaxial, 162
  uniaxial, 118–119
Emmons, R. C., 57, 201
Emmons, S. F., 171
Extinction
  angle of (biaxial), 160–163
  angle of (uniaxial), 117–118
  cause of, 90–91
  inclined, 162–163
  parallel, 117–118, 161–163
  symmetrical, 117–118, 161–163
Extraordinary rays
  definition of, 78
  ray velocity surface of, 79–82
  vibration direction, 80
Eye lens, 20

Faraday, Michael, 7
Fedorow correction diagram for U-stage rotations, 176
Fedorow, E. V., 171, 222
Feldspar
  twinning of, 223
  U-stage studies, 222–230
Field lens, 20
Field-of-view number, 23
Filters
  neutral, 38, 41, 42
  polarizing, 16, 26, 39
Fizeau, Armand H. L., 5
Flash figures
  biaxial, 151
  uniaxial, 113–114
Fletcher, L., 82
Focusing knobs, 35
Foucault, Jean B. L., 1, 5
Fragments, mineral
  advantages of, 231–232
  preparation of, 232
Fraunhofer lines, 48
Frequency of light: *see* Light
Fresnel, Augustin, 1

Gypsum (1st-order red) plate
  construction of, 40
  function of, 99
  use of, 111–112

Hardness, 72
Hemispheres, glass
  for conoscopic light, 175
  universal stage, 172–175
Henry, Joseph, 2
Herapath, William, 16
Hertz, Heinrich, 2
Hornblende
  optical characteristics of, 167–170
  twin, 221
Huygens, Christiaan, 1

Illumination
  adjustment of, 42
  central method, 50–51
  oblique method, 51–53
  for universal stage, 177
  see also Conoscopic illumination, Orthoscopic illumination
Image
  double, 77–78
  imaginary, 18
  principal, 18
  real, 18

Immersion liquids, 47, 59–62
   dispersion of, 48–50, 60–61
   measuring refractive index of, 62–68
   specific gravity of, 60–61
   table of, 60–61
   temperature coefficient of, 58
   use of, 62
Immersion melts, 61
Index of refraction, 10–12, 47
   accuracy of, 56, 58–59
   of biaxial crystals, 122
   definition of, 10
   of isotropic media, 72
   principal, biaxial, 122
   principal, uniaxial, 83, 85
   relation to image depth, 78
   the search for, 114–115, 158–159
   of uniaxial crystals, 80–86
Indicatrix
   biaxial, 121–127
   uniaxial, 82–85
   use of, 85–86, 127–129
Interference colors
   anomalous, 102–103
   chart of (follows 100)
   orders of, 98
   sequence (crossed nicols), 93–98
   sequence (parallel nicols), 98
Interference figures
   biaxial, 136–158, 207–209
   the search for, 114, 158
   uniaxial, 103–114, 196
   see also Dispersion of light
      Flash figures, Optic axis
Iris diaphragm
   illuminator, 41
   objective lenses, 177
   substage, 38, 50
Isochromatic color bands
   biaxial, 136
   uniaxial, 105–106, 112
Isogyres
   biaxial, 136–137
   dispersion of, 154–157
   uniaxial, 105, 106–108
Isotropism, 7, 69–70

Kamb, W. B., 108
Kohler illumination, 42

Land, Edwin H., 16
Light
   addition of, 9
   frequency of, 4
   monochromatic, 50, 59, 61, 97
   nature of, 1–3
   polychromatic, 7
   quanta, 2
   subtraction of, 9
   visible wavelengths, 7
   see also Waves
Lower nicol: *see* Polarizer

Magnification, 31–32
   empty, 32
   total, 18
   useful, 32
Mallard's constant, 137
Maschke, O., 47, 59
Maxwell, James Clerk, 2
Mechanical stage: *see* Stage
Melatope
   biaxial, 136–137
   dispersion of, 155–156
   uniaxial, 105–106
Michelson, Albert A., 5
Michel-Lévy, 17
   method of, 140, 142, 149
Mica ($\frac{1}{4}\lambda$) plate
   construction of, 40–41
   function of, 99
   use of, 112
Microscope
   accessories, 39–41
   adjustment of, 42–46
   development of, 17–18
   function of, 18
   inclined tube, 22, 25
   proper use of, 41–42
   vertical tube, 21, 24
Mirror, substage, 39
Monochromatic light, 50, 59, 61, 97
Moseley, Henry, 2
Muscovite, optical characteristics of, 166–167

Newton, Sir Isaac, 1, 97
Nicol prism, 17, 26–27
   construction and function of, 86–88
   crossed, 46
   parallel, 27, 98
Nicol, William, 17, 26
Numerical aperture, 31–33
   of condensing lens, 37–38
   definition of, 31
   relation to $2V$ angle, 143–144

Oblate spheroid, 80n, 82
Objective lens, 18, 29–35
   achromat-apochromat, 30
   centering of, 35, 42–44
   free working distance, 34, 175
   oil-immersion, 31, 37, 140
   parfocal, 35
   standard markings, 34
   for universal stage, 175, 177

248   *Index*

Ocular (eyepiece) lens, 18, 20, 23
   Huygenian (negative), 20
   Ramsden (positive), 20
Oil, immersion
   for oil-immersion objectives, 32
   for universal stage hemispheres, 175
   see also Immersion liquids
Optic axis
   biaxial, 124, 133
   definition of, 82
   dispersion: *see* Dispersion of light
   interference figures, 108–113, 149–151
   orientation of, 82, 193–195
   uniaxial, 82, 86
Optic normal, 124
Optic plane, 124
Optical angle ($2V$)
   definition of, 125
   dispersion of: *see* Dispersion of light
   measurement of, 137, 140, 142–146
   measurement with universal stage, 207, 209
Optical directions
   biaxial (X,Y,Z), 122
   dispersion of: *see* Dispersion of light
   orientation with universal stage, 205–209
   uniaxial, 116–117
Optical neutrality, 125, 149
Optical orientation
   biaxial crystals, 125, 127, 159–160, 210–215
   monoclinic crystals, 127, 160, 210, 213
   orthorhombic crystals, 127, 159–160, 210
   triclinic crystals, 127, 160, 213, 215
   uniaxial crystals, 116–117, 193–196
Optical sign
   biaxial crystals, 125, 145–151, 207, 209
   uniaxial crystals, 82, 111–113, 195
Ordinary rays
   definition of, 78
   ray velocity surface of, 79–82
   vibration direction, 79–80
Orientation angle, 161, 210–215
Orthoscopic illumination, 103–105, 177, 193, 205

Phase relations of waves, 5, 92–98
Plagioclase, identification of, 222, 226–230
Planck, Max, 2
Plane of polarization
   of analyzer nicol, 26
   of extraordinary waves, 80
   of ordinary waves, 80
   of plane of vibration, 14
   of polarizer nicol, 39
Pleochroism
   of biaxial minerals, 159
   of isotropic minerals, 72
   quantitative and qualitative, 116, 159
   of uniaxial minerals, 115–116
Polarization of atoms, 102, 135–136

Polarized light
   by absorption, 15–16
   definition, 13–14
   by reflection, 14–15
Polarizer (lower nicol), 39
   alignment of, 44–46
   polarization direction of, 39
Polarizing filters: *see* Filters
Polarizing prisms, 26–27, 37, 39, 86–88
Polaroid, 16
Primary colors, 9
Principal sections, indicatrix
   biaxial, 122
   uniaxial, 86
Projections
   gnomonic, 179
   orthographic, 180
   spherical, 179
   stereographic, 180–182
Prolate spheroid, 80n, 82
Pyroxenes
   orientation angle, 213
   twinning, 217

Quartz wedge
   construction of, 40
   function of, 99
   and interference colors, 97–98
   use of, 112–113

Ray, light, 5
Ray velocity surfaces
   in biaxial crystals, 129, 132–133
   definition of, 5, 7
   of extraordinary rays: *see* Extraordinary rays
   in isotropic medium, 70
   of ordinary rays: *see* Ordinary rays
   in uniaxial crystals, 79–82
Reflection
   polarization by, 14–15
   total, 12
Refraction
   angle of, 10–12
   critical angle of, 12
   minimum deviation, 65–68
Refractometer
   Abbe, 62–63
   Leitz-Jelley, 64–65
Relief, 48–49, 51
Resins, cementing, 232–234
Resolution, limit of (resolving power), 31–32
Retardation, 91–98, 100
   addition and subtraction of, 99, 118–119
Roemer, Olaus, 4
Rotation, axes of
   corrections of, 174–175
   universal stage, 171–172
Rutherford, Ernest, 2

Saylor, C. P., 58
Schmidt guide, 175
Schmidt (equal area) net, 181
Sensitive violet, 98
Skiodrome, 137
Slemmons, D. B., 228
Slides
  gelatin-coated, 62, 159
  for petrographic microscope, 233
Snell's law, 10, 78, 79, 137
Sodium vapor lamp, 50
Sorby, Henry Clifton, 17
Specific gravity, 72, 74
  of immersion liquids: *see* Immersion liquids
Stage
  mechanical, 36, 62, 175
  rotating microscope, 36
  universal: *see* Universal stage
Stage micrometer, 23
Stereographic projection, 180–182
  direction of a line, 181–187
  rotation of points, 189–190
  trace of a plane, 188–189
  Wulff net, 180–181
Strain, mechanical, 31, 70

Temperature coefficient: *see*
    Immersion liquids
Thin sections
  advantages of, 232
  preparation of, 232–236
  standard thickness of, 100, 235
Thomson, J. J., 2
Topaz, optical characteristics of, 164
Tourmaline, absorption of, 15–16
Triaxial ellipsoid, 122–123
Trichroism, 159
Triphylite, optical characteristics of, 164–165
Tube, microscope, 18, 20
  length, 31, 34
Turner, F. J., 228
Twin axis, 197

Twin composition plane, 197
Twin plane, 197
Twinning
  of biaxial crystals, 215–222
  of feldspars, 223
  of monoclinic crystals, 217
  of orthorhombic crystals, 217
  of triclinic crystals, 222
  of uniaxial crystals, 197–201
Twins
  complex, 216
  contact, 197
  normal, 197, 215–216
  parallel, 215–216
  penetration, 197
  polysynthetic, 197
  repeated, 197
  simple, 197
  symmetrical (cyclic), 197

Universal stage
  axes of, 171–172
  conoscopic illumination, 196, 207, 209
  glass hemispheres: *see* Hemispheres
  mounting and adjusting, 177–178
  orthoscopic illumination, 193–195, 205–207
Upper nicol: *see* Analyzer

Velocity of light, 4–5

Waves
  electromagnetic, 2, 7, 13–14
  longitudinal, 1
  motion of, 3–4
  transverse, 3–10
Wavelength
  of light, 4, 7
  of monochromatic sodium light (N_D), 50
Wright arcs, 172, 183
Wulff (stereographic) net, 180–181

Young, Thomas, 1